Comparative Cardiovascular Dynamics of Mammals

Comparative Cardiovascular Dynamics of Mammals

John K-J. Li

CRC PRESS

Boca Raton   London   New York   Washington, D.C.

Library of Congress Cataloging-in-Publication Data

Li, John K-J., 1950-
  Comparative cardiovascular dynamics of mammals / John K-J Li.
        p.   cm.
  Includes bibliographical references and index.
  ISBN 0-8493-0169-6 (alk. paper)
  1. Hemodynamics. 2. Physiology, Comparative. I. Title.
  [DNLM: 1. Cardiovascular System—physiology. 2. Mammals—physiology. 3. Anatomy,
Comparative—methods. 4. Hemodynamics—physiology. 5. Models, Cardiovascular. WG
102L693c   1996]
QP105.L5   1996
599'.011—dc20   95-45491

Visit the CRC Press Web site at www.crcpress.com

© 1996 by CRC Press LLC

No claim to original U.S. Government works
International Standard Book Number 0-8493-0169-6
Library of Congress Card Number 95-45491

Dedication

To those who have brought happiness to my life,
especially my sons Michael and Christopher.

Preface

Establishment of similarity principles is a necessary and important step in understanding the natural common design features of the cardiovascular system of different mammals. Combination of the classic allometric method, dimensional analysis, and modern hemodynamic principles provides a more meaningful and a more powerful approach to obtain new similarity principles. Relating local fluid and blood vessel mechanical properties to global pulse transmission characteristics enables identification of arterial system similarities. Similarity principles derived for the heart may reveal the functional design criteria based on mammalian cardiac energetics; its force development, the energy generation, and their relation to external mechanical work and body metabolism during each heart beat. Analyzing the heart-arterial system interaction allows understanding of how the energetic muscular pump is coupled to its pulsating vascular load system. Extraction of measured data and parameters from well-designed animal experiments are useful to elucidate established similarity principles. Evaluation of the sensitivity of parameters and similarity principles may provide an indication as to whether these principles can be employed as potential tools in the diagnosis of cardiovascular diseases. The goal of this book is to provide an insight to the following questions: Is the natural mammalian cardiovascular system an optimal system? If it is, then what are the optimal design features? Why are the structural design features closely coupled to hemodynamic functions? And finally, are these functions governed by similarity laws? Answers to these questions may advance our current understanding as to the natural processes involved in the functional development of the mammalian cardiovascular system.

The Author

John K-J. Li, Ph.D., is Professor of Biomedical Engineering and Director of Cardiovascular Research at Rutgers University, New Jersey.

Dr. Li obtained a B.Sc. honors degree from the University of Manchester, England, in 1972, and the M.S.Eng. and Ph.D. degrees in bioengineering from the University of Pennsylvania in 1974 and 1978, respectively. He was Head of Biomedical Engineering in Cardiology at Presbyterian-University of Pennsylvania Medical Center in Philadelphia from 1977 to 1979. He joined Rutgers University as an Assistant Professor in 1979, became an Associate Professor in 1983 and a Professor in 1989. He is also an Adjunct Professor of Surgery of the UMDNJ-Robert Wood Johnson Medical School.

Dr. Li is a Fellow of the American College of Cardiology, a Fellow of the American College of Angiology, and a member of the American Heart Association, Cardiovascular System Dynamics Society, American Physiological Society, American Society of Hypertension, Biomedical Engineering Society, and the Engineering in Medicine and Biology Society of the IEEE. He has been an Editorial Board member and a reviewer of several leading journals. He has received awards and certificates from Rutgers, the IEEE, and the International Union of Physiological Congress. He has been a recipient of research grants from the National Science Foundation, the National Institutes of Health, and the American Heart Association. He is a frequently invited speaker and scientific session chairman in national and international cardiovascular and biomedical engineering conferences.

Dr. Li is the author of more than 300 publications. He is also the author of the book entitled *Arterial System Dynamics*, and co-editor of two conference proceedings. His current research interests include cardiovascular dynamics, hypertension, myocardial ischemia, comparative physiology, innovative biomedical instrumentation, and cardiac assist devices.

Table of Contents

Chapter 9
Optimality and Similarity **137**

CHAPTER 1

Introduction

1.1 THE MAMMALIAN SPECIES

Apart from creation, evolution has become a science that provides an interpretation of the general view of the existence of the mammalian species and an understanding of their bodily functions. According to evolutional biology, reptiles dominated the world millions and millions of years ago. Mammals were evolved from primitive reptiles called pelycosaurs. Small mammals appeared some 150 million years ago. The main purpose of this book is to present a scientific treatise on the cardiovascular function of mammalian species existing in the present age or Pleistocene mammals. This book makes no attempt to deal with evolution or to support the evolution theory of Charles Darwin.

The mammalian species considered here include *Homo sapiens*, or humans; domestic and wild mammals; and marsupials, such as kangaroos, wombats, and bats, the only flying mammals. Sea mammals, such as the dolphin and whale, are beyond the scope of the present coverage. The marsupials exist largely in Australia, a continent that also has very few (but imported) elephants, the largest existing land mammal. The largest land mammal ever known to have existed was Baluchitherium, which was said to have a body weight over 30 tons. This terrestrial mammal is an ancestral relative of the rhinoceros. From the general observation that the larger the body size of a mammal, the slower is its heart rate, Baluchitherium could have had a heart rate of about 15 beats per minute. A long cardiac cycle of 4 sec. was obviously required to circulate enough blood to perfuse the vast organ vascular beds.

The mammalian species are distinguished from one another immediately by body size and physical features. Although intraspecies differences are minimized, the growth patterns show considerable differences between infants and adults. In the young, the heart is small in size with a higher heart rate. Individual organs also show differential growth in relation to body size. The differential growth is perhaps a result of the complimentary correlates of structure and function.

The characteristics of the mammalian species can be explored through allometric extrapolations and modeling predictions of common features. Both are included in the current treatment and methodological approaches. Analysis will be based on available cardiovascular measurements on selected species.

1.2 THE CARDIOVASCULAR SYSTEM

Structural design features that complement the function of the cardiovascular system in man have intrigued researchers for centuries. To trace the development of the understanding of the anatomical structure and physiological function of the system is of great interest here. Indeed, the pulsatile nature of blood flow was not known until William Harvey (1626), although pulsations had already been described by Chinese practitioners in ancient times. Harvey, in his well known book *De Motu Cordis* described the intermittent nature of the pumping heart. This leads to the definition of a systole as when the heart contracts to eject blood and diastole, then the heart relaxes and is filled with the returning blood. The heart does this with its constituent mucle fibers, thus, it is known as a muscular pump.

The mammalian heart has four chambers. The right and left ventricles serve as synchronized pumps which eject blood to the pulmonary and the systemic arterial trees and through arterioles to capillaries. In this manner, vital organs and their vascular beds are perfused. Small venules and veins return blood to the left and right atria. Figure 1.1 illustrates the circuits.

Mammalian hearts have two ventricles. In this respect, they differ from, for example, amphibian hearts, which have only one ventricle called an univentricular pump. What are the advantages of dual ventricular pumps? It has been suggested that evolutionary processes are involved in the functional demand of oxygenation by warm-blooded mammals (Longmore, 1972). It is interesting to note that, even at this stage, structure-function complement is an important part of the natural design features. The two ventricles are synchronized during cardiac pumping. Although blood pressures are different in the pulmonary and the systemic arteries, the volume ejected during each contraction, or the stroke volume, is the same for the two ventricles, since the same amount of blood is circulated.

Some excerpts from earlier cardiovascular studies may be of interest to readers. Ancient Chinese physicians had already felt palpable pulsations in radial arteries as a means of diagnosing the state of the heart. This implies that the transmission of pulsations from the heartbeat was already known. It was said that Herophylos of Alexandria in the fourth century B.C. also noted that the heart transmits blood and pulsations to the arteries. However, he interchanged the usage of arteries and veins, i.e., he saw the pulmonary artery as the arterial vein and the pulmonary vein as the venal artery. Erasistratos apparently was the first to give a detailed description of the heart valves, both tricuspid and aortic, which were later found by Galen to be one-way valves. Galen apparently realized the important aspect of the circulatory function of oxygen (known only as air in his time) transport and the removal of waste products.

Harvey, mentioned earlier, effectively defined systole as the contraction of the heart and diastole as relaxation of the heart to fill with blood. Lower described the

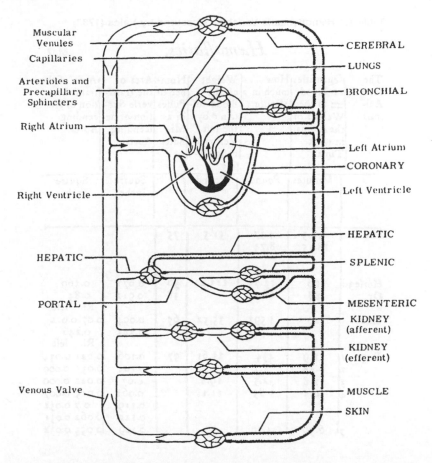

Figure 1.1 Block diagram showing the principal circuit of blood flow of the mammalian cardio-vascular system. (From Selkurt, E. E., *Physiology*, Little, Brown, Boston, 1971. With permission.)

contraction of the heart when the apex comes closer to the base, or the long axis shortening. This movement is correctly described and can be observed with modern-day X-ray angiograph or ultrasound echocardiography. His dissection of eight animal species included both mammals and amphibians. It is clear that the hemodynamic differences between the two was not yet established at that time. Galileo, interestingly, was his teacher at one time.

Stephen Hales, in *Statical Essays Containing Haemostatics* (1733), provided the first quantitative examination of the circulation, utilizing "cardiovascular instruments" for the measurement of blood flow and blood pressure. Hales' quantitative account of the output of the heart was his greatest contribution, especially the concept of the blood pressure and flow and their interrelationship. These pioneering measurements formed an important part of modern day hemodynamics. From the standpoint of comparative hemodynamics of mammals, Hales already began such analysis in the 18th century. Table 1.1 lists some of the cardiovascular parameters he obtained in

Table 1.1 Hemodynamic Variables Measured by Hales (1733)

Hæmaſtatics.   43

The several Animals.	Quantities of Blood == to the Weight of the Animal in what Time.	How much in a Minute.	Weight of the Blood suftain'd by the left Ventricle contract-ing.	Num-ber of Pulses in a Mi-nute.	Area of the tranf-verfe Sec-tion of de-fcending Aorta.	Area of the tranf. Sec-tion of af-cending Aorta.
	Minutes	Pounds	Pounds		Square Inches	Square Inches
Man	36.3 18.15	4.37 8.74	51.5	75		
Horfe 3d	60	13.75	113.22	36	0.677	0.369
Ox	88	18.14		38	0.912	0.85 Ri. left
Sheep	20	4.593	35.52	65	0.094 0.383	0.07 0.012 0.246 Ri. left
Dog 1	11.9	434	33.61	97	0.106	0.041 0.034
2	6.48	3.7			0.102	0.031 0.009
3	7.8	2.3	19.8		0.07	0.022 0.009
4	6.2	1.85	11.1		0.061	0.015 0.007
					0.119	0.7 0.031
					0.125	0.062 0.031
7	6.56	4.19			0.109	0.053 0.032

man, horse, ox, sheep, and dog and include the quantity of blood (blood volume), cardiac output, heart rate, ascending and descending aortic cross-sectional areas. These latter two measurements, as we shall see later, illustrate the effects of "geometric taper". Some of his measurements, though crude by present-day high-technology standards, nevertheless gave very good quantitative estimates. They are also not too far from those predictable from corresponding allometric equations, as we shall see later.

1.3 FUNCTION OF THE CIRCULATION

The principal function of blood is to maintain a constant cellular environment by circulating and delivering nutrients to body organ tissues and removing waste products from them. Blood itself consists of plasma fluid and formed elements, such as red blood cells (erythrocytes), white blood cells (leukocytes), and thrombocytes (platelets). The density of blood at 37°C is about 1.06 g/cm^3. The viscosity at corresponding temperature is about 3 cP or three times more viscous than water.

Plasma contains many important components such as proteins, carbohydrates, electrolytes, lipids, and hormones. The proteins include globulin, albumin, and fibrinogen.

The main function of red blood cells (RBC) is to transport oxygen and carbon dioxide. In man, there are some 4.5 to 6 million RBC/ml of blood. The relative content of RBC in blood determines the hematocrit,

$$\text{Hematocrit } (\%) = (\text{RBC/Blood}) \times 100\% \qquad (1.1)$$

This percentage is higher in venous blood than arterial blood and varies among tissues. For instance, it is about 45% in veins, 20% in kidneys, and 70% in splenic circulation.

The main function of white blood cells (WBC) is to protect against the invasion of infection and bacteria. These cells can pass through the vascular endothelium and enter the tissue spaces, depending on the need to destroy undesirable foreign substances by phagocytosis or other cellular movement. In man, there are about 5000 to 7000 WBC per milliliter of blood. Platelets, on the other hand, play a major role in coagulation. There are also circulating catecholamines that contribute to the body's ability to meet hemodynamic functional demands. Epinephrine and norepinephrine are examples.

Blood circulation is vital to body organ functions. It not only meets metabolic demand, but also maintains body temperature. This is well illustrated by the movement of arteries and veins operating under the countercurrent mechanism.

1.3.1 MAMMALIAN CELL SIZE AND TISSUE METABOLISM

Mammals, large and small, have their fundamental building blocks, the cells, that are roughly the same size, within an order of magnitude of 10 μm. However, the muscular mitochondrial density is higher in smaller mammals. This has been attributed to a higher metabolic turnover rate, as we shall discuss later.

Red blood cells in various mammals are also of much the same size. In fact, they are independent of body size. In general, a large mammal is not made up of larger cells, but of a larger number of cells. The vascular bed in a large mammal therefore contains a large number of capillaries and capillary networks with a large number of red blood cells of the same size.

Various organs have considerably different metabolic rates and hence different blood perfusion rate. The most active organs, such as the kidney, the heart, and the brain, have metabolic rates over 100 times those of less active organs, such as skin, bone, and adipose tissue. Thus, metabolic activity is coupled to blood flow and oxygen consumption. The tissues are normally supplied with glucose at a rate corresponding to the use of O_2.

The observed decrease in specific metabolic rate cannot be explained by decreases in the relative sizes of the metabolically most active organs. This, as we shall see in a later chapter is intimately related to heart rate.

Oxygen transport to meet metabolic demand in various mammals is dependent on the concentration of hemoglobin, the total volume of blood, the number of red blood

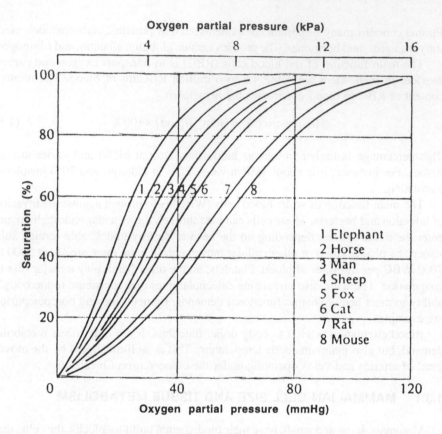

Figure 1.2 Oxygen dissociation curves for the blood in different mammals. (From Schmidt-Nielsen, K., *Scaling: Why is Animal Size so Important?*, Cambridge University Press, 1984. With permission.)

cells, and the affinity of hemoglobin for oxygen. The oxygen-carrying capacity of the blood is directly proportional to its hemoglobin concentration. The average hemoglobin concentration in different mammals has been reported to be about 130 g/L of blood. The corresponding mean oxygen capacity of the blood is about 175 ml/L of blood. This latter value was found to be relatively constant in mammals of different body sizes.

Numerous measurements have shown that in terrestrial mammals, blood volume is a constant fraction of body weight, while the hemoglobin concentration of the blood is independent of body size. The total amount of hemoglobin present in the body available for oxygen transport is therefore a constant fraction of the body weight. The binding between hemoglobin and oxygen, or the so-called oxygen dissociation curve, describes the degree of saturation of the hemoglobin with oxygen at any given oxygen concentration. A number of typical oxygen dissociation curves for mammalian blood are shown in Figure 1.2. Spectrophotometric methods are commonly used for the determination of hemoglobin saturation and organ oxygenation.

If we compare the sizes of red blood cells from various mammals, we find the surprising fact that their diameters seem to be rather uniform and, thus, independent of body size. Extensive data on red cell size are available (Altman and Dittmer, 1961). A survey of red blood cells from more than 100 species of mammals show that most of the diameters are in the range of 5 to 8 μm, without any correlation with body size. The smallest mammal, the shrew, and one of the largest, the humpback whale, have red cells of very similar sizes, 7.5 and 8.2 μm, respectively. No mammal has a red blood cell diameter over 10 μm, and only a few (sheep, goat, deer) have red cell diameters less than 5 μm.

1.4 BOOK CONTENT

Similarity of objects must have existed since life began. The conceptual analysis of similarity, however, was not formulated until modern civilization.

Mathematically similar systems were demonstrated by the ancient Chinese and Greeks through geometry. Physically similar objects were reasoned by Galilei (1638), who discovered mechanically similar objects, by Newton (1735), whose laws of motion are perhaps the best-known laws of physics, by Reynolds (1883), who discovered the dimensionless Reynolds number in fluid mechanics, and by others. Biological similarity analyses lagged behind those of mathematics and physics for many centuries. Leonardo da Vinci's paintings of human nudes suggest his conceptual ideas of human structural similarities. The use of modern similarity and dimensional analysis in biological, particularly physiological systems, are relatively recent and remains scant.

In the circulatory system, it appears that dimensional analysis was applied only in the last 50 years. It is surprising that similarity analysis accounts for only a small portion of the vast amount of cardiovascular research, where a great amount of experimental data is available.

In this book, I shall review the foundation of biological similarity analysis, specifically where such analysis is related to the hemodynamics of mammals. I shall apply this tool to identify similarity principles that are fundamental to the common function of the mammalian cardiovascular system.

This book takes a new approach that embraces the classical allometric method, dimensional analysis, and modern cardiovascular dynamics to establish new similarity principles for the arterial system, the heart, and the cardiovascular system in a closed loop. Analyses are performed on selected mammalian species but always with reference to their respective body masses. The sensitivity of these similarity criteria during altered pathophysiologic conditions is examined. Analysis is also applied to identify optimalities in the natural mammalian cardiovascular system and to reveal the corresponding similarities in exhibiting optimal design features of the structure and function of the cardiovascular systems of different mammals.

In this chapter, the evolution of mammals with their general features, the components of the cardiovascular system, and the physiologic function of the circulatory system are discussed.

In Chapter 2, we shall take a more detailed look at the common features of the anatomic structure and the functional physiology of the individual components of the mammalian cardiovascular system, in terms of their relative geometric proportions and their physical behaviors. The heart acts as the sole energy source, with the arteries acting as its delivery conduits to the perfused microvascular beds, and the veins acting as the collecting ducts and the reservoirs of the system.

In Chapter 3, the basic physical principles of comparing different mammalian species will be presented. These include, for instance, the use of modeling and geometric ratios, the derivation and utilization of the powerful allometric equation, and the quantitative description of physiological growth and differential growth of organs and organisms. How these types of growth lead to the proportionate or disproportionate changes in organs relative to body size will be addressed, as well as the importance of body weight. The use of allometric equations to describe cardio-vascular measurements and observations will be demonstrated. Finally, use of the principle of allometry to quantify the dynamics of blood flow and cardiac function will be elucidated.

Chapter 4 begins by building the mathematical foundations for readers, or serves as a review for those already familiar with the mathematical concepts. These foundations include the logarithmic relationships and exponents, linear algebra through matrices and determinants, and the basic calculus through linear differential equations. The treatment is relatively basic such that readers can follow the subsequent sections of the chapter. The dimensional analysis utilizes the mass (M), length (L), and time (T) system, and the formation of dimensional matrices based on selected hemodynamic parameters that are considered pertinent. Buckingham's Pi theorem is used to derive pi-numbers, which are dimensionless. The similarity principles are established when the pi-numbers, or the combinations of pi-numbers, are also inde-pendent of mammalian body weight, i.e., as invariants. Illustrative examples will be provided here, although more elaborate treatment can be found in chapters that follow.

Chapter 5 deals with the mechanics of the heart. Sarcomeres are the basic building units of the ventricles. The following questions are answered. How does each individual myofibril contribute to the pumping action of the heart? How is the input-output relationship dependent on the operation of the Frank-Starling mecha-nism? How does the atrial filling pressure influence the performance of the ven-tricles? How is the muscle contractility determined in ejecting hearts? Why is the fraction of blood ejected out the ventricle about the same throughout mammalian species? The modern treatment of the pressure-volume relationship of the heart from a revived old concept is used to describe global cardiac function.

What follows naturally from the heart is the arterial system. Here we deal with the mechanics of blood vessels, the interplay of the geometric and elastic properties that leads to the manifestation of similar blood pressure and flow waveforms in corresponding anatomic sites of different mammals, and the concept of vascular-input impedance. Salient features of the propagation and reflection of the pulses and of the heart-generated energy and its dissipation en route to organ vascular beds are explained.

Having elucidated the dynamics of the system, a few examples of the similarity analysis embodying such understanding are presented as selected topics. These include the fluid mechanical aspects of laminar and turbulent flows, the tension-length relationship and Laplace's law, the heart rate mystery (why hearts of small mammals beat faster), the efficiency of the heart as a mechanical pump, the microvasculature, and the significance of the interplay of heart rate, energy, and metabolic demand in terms of myocardial oxygen consumption for the heart and metabolic turnover rate for the body.

The final two chapters deal with advanced methods and applications of similarity analysis, in particular, how the similar features are inherent among species, and whether they represent optimal structure and functional correlates, and how such optimality can be identified.

CHAPTER 2

Comparative Anatomy and Physiology of the Circulation

2.1 THE HEART

2.1.1 THE EVOLUTION OF THE MAMMALIAN HEART

It has been hypothesized that about 450 million years ago, the common ancestor of all fishes, amphibians, reptiles, birds, and mammals already had a serially connected four-chambered heart. In this hypothesis, evolution of the heart began with the fish some 400 million years ago followed by amphibians some 300 million years ago, and lastly, the mammalian and avian species some 200 million years ago. This was illustrated by Longmore (1972), as shown in Figure 2.1.

About 220 million years ago, certain members of the class Reptilia developed a true four-chambered heart. A septum was present that divided the ventricle into left and right chambers. Crocodiles are the only known surviving reptiles with four-chamber hearts. However, the complete separation of oxygenated and deoxygenated blood exists only in mammals and birds. This development and separation of systemic arteries and veins allows regulation of body temperature and the maintenance of core temperature to within fractions of 37°C. The countercurrent exchange mechanism operating between the arteries and the veins contributes to peripheral temperature regulation.

The modern parallel arrangement of atria and ventricles offers less resistance to blood flow. This stems from the basic fact that serial resistances add. The dual enlarged ventricles that are synchronized during pumping generate more power and make it possible for the heart to pump a larger volume of blood than is possible with the rudimentary serial connections.

Similar anatomic features among mammalian circulatory systems have long been recognized. The mammalian heart has four chambers, the left and right ventricles and the left and right atria. The atria are thin walled (small radius-to-wall-thickness ratio) chambers, and the two ventricles are thick-walled chambers. Each of the atria also has an appendage; the left atrial appendage and the right atrial

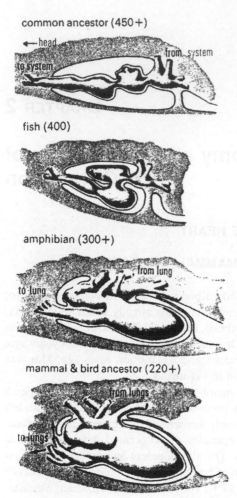

common ancestor (450+)

fish (400)

amphibian (300+)

mammal & bird ancestor (220+)

Figure 2.1 Serially connected four chambers of the heart. (From Longmore, D., *The Heart*, McGraw-Hill, New York, 1971. With permission.)

appendage (Figure 2.2). The left ventricle pumps blood into the aorta and the right ventricle into the main pulmonary trunk.

The atrial appendages have no specific functions apart from those of the atria; within a closed pericardial space they are filled with blood during atrial diastole and emptied into the ventricle during atrial systole.

As mentioned earlier, the atrial muscles are thin and able to develop only relatively low pressures when they contract. The right atrial wall tends to be somewhat thinner than the left, and the right atrial pressure is consequently slightly lower than the left. This is also seen as a consequence of Laplace's law (Chapter 7). The atrial chambers do not normally contribute much to resting cardiac function. Though debatable, the atrial filling has generally been considered a passive process.

The shape of the left ventricle is between conical and semiellipsoidal with its narrow end, the apex, forming the apex of the heart. Both shapes, as well as the cylinder and sphere have been used in ventricular modeling. The left ventricular wall

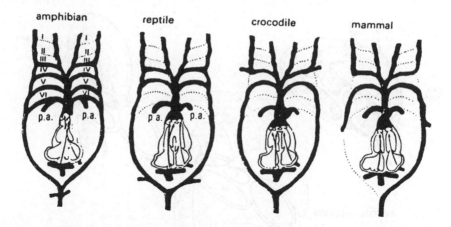

Figure 2.2   Anatomical organization of the mammalian heart, showing four chambers. (From Longmore, D., *The Heart*, McGraw-Hill, New York, 1971. With permission.)

is about three times as thick as the right ventricular wall; thus it is able to develop a much higher pressure. Again this is a consequence of Laplace's law. The wall is thickest around the circumference, or the widest part, of the cavity and thinnest at the apex. In a practical sense, the thick interventricular septum is more closely associated with the pumping action of the left ventricle. The ventricle also shortens much more in the short-axis, or circumferential, direction than in the long-axis, or base-to-apex, direction.

The right ventricle resembles a section of a sphere (Figure 2.3), with the left and right ventricles separated by the interventricular septum. The free (outer) wall of this sphere is pulled toward the interventricular septum by contraction of the thick left ventricular transverse fibers, to which the right ventricle is attached. The left ventricle is said to assist the pumping action of the right ventricle because its wall has greater mass and a greater contractile force. The extent of ventricular interference, or cross talk, has been a subject of many recent investigations. In general, the right ventricle, with its thin wall, is constructed for pumping blood at low pressures that exist in pulmonary circulation. The thicker left ventricle is better suited to pumping into the higher-pressure systemic circulation. The ventricles are made up of muscular fibers. This so-called myocardium can be further divided transmurally into the inner endocardium and the outer epicardium.

There are four heart valves involved in the filling and pumping action of the heart (Figure 2.4). The mitral valve, situated between the left atrium and the left ventricle, controls the flow between these two chambers, but is a one-way valve. The mitral valve opening facilitates inflow from the left atrium into the left ventricle. The mitral valve has a large anterior cusp and a small semilunar posterior cusp. The anterior leaflet of the mitral valve forms one boundary of both the inflow tract and the outflow tract of the left ventricle. During diastolic filling of the left ventricle, the anterior leaflet of the mitral valve swings forward toward the septal wall to provide maximum movement, allowing blood from the left atrium into the left ventricle. During left

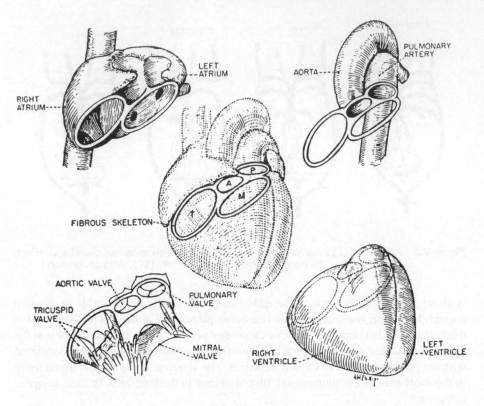

Figure 2.3   Sketch showing the geometric relations of the left ventricle (LV), the right ventricle
(RV), and the interventricular septum (IVS). (From Rushmer, R. F., *Structure and
Function of the Cardiovascular System*, W. B. Saunders, Philadelphia, 1972. With
permission.)

ventricular ejection of blood, the anterior leaflet of the mitral valve in its closed
position forms the posterior wall of the outflow tract of the left ventricle. These
motions are easily seen in ultrasonic M(motion)-mode scan images, frequently used
in clinical echocardiography.

The tricuspid valve, as the name implies, has three cusps. Named the posterior,
the septal, and the anterior, the cusps have similar geometric shapes.

The right ventricle and the low-pressure pulmonary arterial system, on the other
hand, is separated by the pulmonic valve. The aortic valve separates the left ventricle
from the ascending aorta, which leads to the high-pressure systemic arterial system.
These valves are tricuspid valves whose leaflets are of similar shape. The function
of these valves in relation to the functional dynamics of the cardiovascular system
will be incoporated later in our studies of the heart.

The heart is enclosed in a pericardial sac, which is both fibrous and serous. Under
normal physiological conditions, the pericardial sac contains only a few milliliters of
fluid. An excess amount of pericardial fluid can lead to severe hemodynamic consequences in various disease states, such as cardiac tamponade.

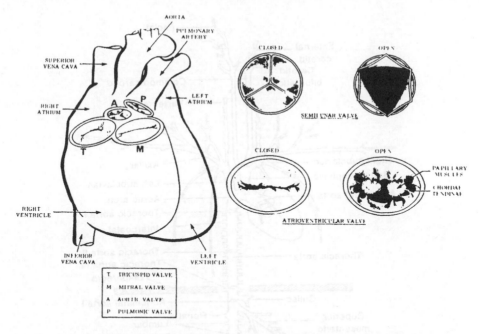

Figure 2.4   Diagram showing the positions of the four cardiac valves. (From Rushmer, R. F., *Structure and Function of the Cardiovascular System*, W. B. Saunders, Philadelphia, 1972. With permission.)

2.2 ARTERIES

2.2.1 ANATOMICAL AND STRUCTURAL ORGANIZATION OF THE ARTERIAL TREE

Detailed descriptions of the mammalian arterial tree can be found in many anatomical textbooks. The major branches of the arterial tree are shown in Figure 2.5. The aorta forms the outflow tract of the left ventricle. The root of the aorta begins immediately at the aortic valve. The aorta features an arch. The aortic arch junction is formed by the ascending aorta, the brachiocephalic artery, the left subclavian artery, and the descending thoracic aorta. Distal to the arch, the aorta continues to its descending portion, which has many branches perpendicularly connected to supply blood via renal arteries to the kidneys and via other arteries to their respective organs. Its distal end is the abdominal aorta. Main branching (bi- or trifurcation) appears at the end of the abdominal aorta (aorto-iliac junction). This junction is formed by the abdominal aorta and the left and right iliac arteries and its continuation. In the human, it is a bifurcation.

The femoral artery, the best known peripheral artery because of its accessibility, continues from the iliac artery to the tibial artery. The common carotid arteries are the longest geometrically uniform vessels with the least geometrical tapering. The arterial system is otherwise seen to be a tapered branching system. Changes in lumen

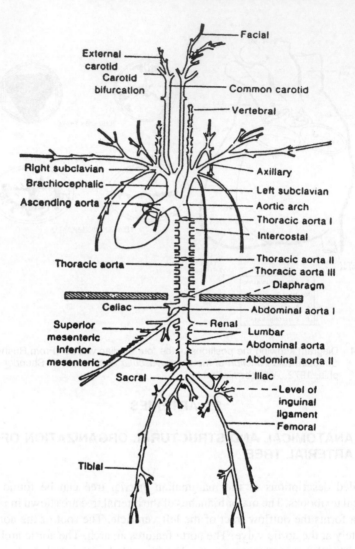

Figure 2.5 An illustrative example of the major branches of the mammalian arterial tree. (From McDonald, D. A., *Blood Flow in Arteries*, Arnolds, London, 1960. With permission.)

size are associated with branching and appropriate tapering. In the normal arterial system, the branched daughter vessels are always narrower than the mother vessel, but with slightly larger total cross-sectional areas. The area ratio, the ratio of the sum of the daughter vessels' cross-sectional areas to the mother vessel's cross-sectional area, is slightly greater than one.

The luminal cross-sectional area decreases progressively from large arteries to smaller peripheral vessels. The total cross-sectional area also increases at vascular branching. The increased cross-sectional area facilitates perfusion of organ vascular beds.

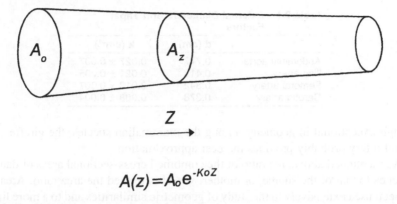

$$A(z) = A_o e^{-koz}$$

Figure 2.6 Geometric taper of a blood vessel.

Arterial diameters and lumen cross-sectional areas of the vascular tree can be determined from postmortem cast or angioradiography. The number of generations of blood vessels for different mammals is given by Green (1950).

As mentioned above, there is appreciable "geometric taper" in the aorta and smaller but noticeable taper in its smaller arteries. Together with branching, it contributes to the "geometric nonuniformity" observed throughout the arterial system. The term *geometric taper* is appropriate only for a single, continuous conduit, such as the aorta. This taper effect is shown in Figure 2.6. The area change of the aortic cross section closely follows an exponential form and can be expressed as

$$A(z) = A_o e^{-kz/r} \qquad (2.1)$$

where k is the geometric taper factor

$$k = r/z \ln(A_o/A) \qquad (2.2)$$

A_o is the lumen area at the proximal site, r is the internal or luminal radius, and z is the distance in the longitudinal axial direction. Li (1987) has reported a value of k in the aorta between 0.031 to 0.037 for dogs between 20 to 30 kg. Here k is a constant. Although geometric taper in different mammals has not been extensively examined, such taper factor is expected to be similar in other mammalian aortas.

Axial taper factors k_o (cm^{-1}) for blood vessels can be computed from the formula:

$$A = A_o e^{-k_o z} \qquad (2.3)$$

The reported values of k_o for some vessels (Li, 1978) are shown in Table 2.1. These are measured *in vivo* in experimental dogs at a mean arterial pressure of about 90 mmHg and average body weight of about 20 kg. It is clear from this table that the taper factor is smaller for smaller vessels. Carotid arteries have the least taper, and they are thus a better approximation to a geometrically uniform cylindrical vessel.

Table 2.1   External Diameters and Taper
            Factors

	d (cm)	k (cm⁻¹)
Abdominal aorta	0.777	0.027 ± 0.007
Iliac artery	0.413	0.021 ± 0.005
Femoral artery	0.342	0.018 ± 0.007
Carotid artery	0.378	0.008 ± 0.004

Though exceptional in geometry among the mammalian species, the giraffe's long carotid artery probably provides the best approximation.

As mentioned above, the ratio of the combined cross-sectional areas of daughter branches to that of the source, or mother, vessel is termed the area ratio. Area ratio has been used extensively in the study of geometric similarities and to a more limited scope, in hemodynamic studies. For most mammalian aortas, the area ratio is slightly higher. In the dog it is about 1.08 at the aortic arch, and 1.05 at the aorto-iliac junction (Li et al., 1984). Deviations from the norms will produce an adverse effect on cardiovascular function. This aspect is dealt with in Chapter 9.

In the broadest sense, the arterial wall (Figure 2.7) consists of three main components, the elastin, collagen, and smooth muscle embedded in a mucopolysaccharide ground substance. A cross section of an artery reveals the tunica intima, which is the innermost layer consisting of a thin layer (0.5 to 1 μm) of endothelial cells, connective tissue, and basement membrane. All blood vessels, including the capillaries, have endothelium.

The middle layer of the arterial wall is the thick tunica media, separated from the intima by a prominent layer of elastic tissue, the internal lamina. Elastic laminae are concentrically distributed and attached by smooth muscle cells and connective tissue. The media contains elastin, smooth muscle cells, and collagen fibers. The difference in their composition divides arteries into elastic and muscular vessels. The aorta is considered an elastic vessel, while the small femoral artery belongs to the muscular vessels. The relative content of these components in different vessels is shown in Figure 2.7.

The outermost layer is the adventitia, which is made up mostly of stiff collagenous fibers. The high elastic stiffness of the collagen fibers is known to prevent blood vessels from rupture under high pulsating blood pressures.

Longitudinally, we find that the number of elastic laminae decreases with increasing distance from the aorta, but the amount of smooth muscle increases and the relative wall thickness increases. The net stiffness is also increased, accounting for the increase in pulse wave propagation velocity. The mechanical behavior of peripheral vessels is largely influenced by the behavior of the smooth muscle, particularly by its degree of activation.

2.3 THE VEINS

Arteries deliver blood from the ventricles to vascular beds, while veins return it to the atria. Veins, unlike arteries are generally thin-walled and have low distending

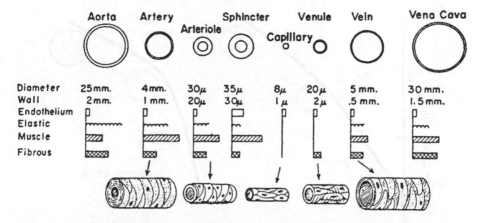

Figure 2.7 Relative contents of elastin, collagen, and smooth muscle in blood vessel walls. (From Rushmer, R. F., *Structure and Function of the Cardiovascular System*, W. B. Saunders, Philadelphia, 1972. With permission.)

pressures. They are collapsible even under normal conditions of blood pressure swing.

The inferior vena cava is the main trunk vein. The superior vena cava feeds into the right atrium, and the main pulmonary vein leads into the left atrium with oxygen-enriched blood.

The body has a greater total number of veins than arteries, and thus the venous system has a much larger cross-sectional area. This results in a much larger volume available for blood storage. Indeed, veins are known as low-pressure storage reservoirs of blood. Under normal physiological conditions, the venous system contains about 75% of the total blood volume in the systemic circulation, with the systemic arterial system constituting some 15%. For this reason, veins are often referred to as capacitance vessels. Venous return is an important determinant of cardiac output. The pulmonary circulation contains about one quarter the blood volume of the systemic circulation.

Veins have much thinner walls and fewer elastins than arteries. Because of this, veins are stiffer than arteries. However, the low operating pressure and collapsibility allow veins to increase in volume by several times under a small increase of distending pressure. This is illustrated in Figure 2.8.

There are bicuspid valves in veins (Figure 2.9). These valves permit unidirectional flow, thus preventing retrograde blood flow to tissues due to high hydrostatic pressures. These valves are notably present in the muscular lower limbs.

2.4 THE MICROVASCULATURE

As stated previously, the function of the cardiovascular system is to provide a homeostatic environment for the cells of the organism. The exchange of the essential

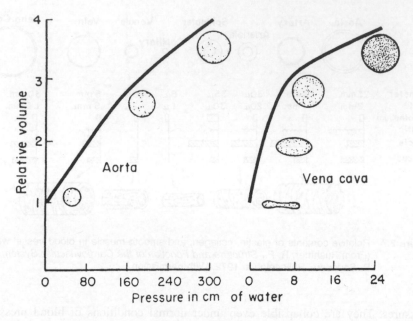

Figure 2.8 Comparison of pressure-volume curves and cross-sectional areas of the aorta and vena cava. (From Little, R. C., *Physiology of the Heart and Circulation*, Year Book Medical Publishers, Chicago, 1985, p. 209. With permission.)

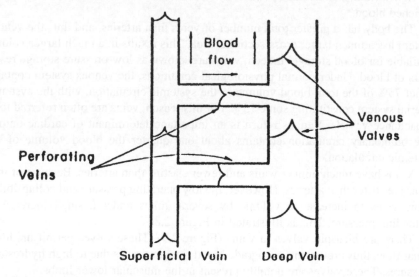

Figure 2.9 Bicuspid valves in a segment of vein.

nutrients and gaseous materials occurs in the microcirculation at the level of the capillaries. These microvessels are of extreme importance for the maintenance of a balanced constant cellular environment. Capillaries and venules are known as ex-

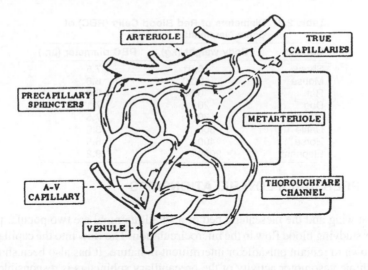

Figure 2.10 A network of the microcirculatory unit. (From Selkurt, E., *Physiology*, Little, Brown, Boston, 1971. With permission.)

change vessels where the interchange between the contents in these walls and the interstitial space occurs across their walls.

2.4.1 ANATOMIC ORGANIZATION

The microcirculation can be described in terms of a network such as that shown in Figure 2.10. It consists of an arteriole and its major branches, the metarterioles. The metarterioles lead to the true capillaries via a precapillary sphincter. The capillaries gather to form small venules, which in turn become the collecting venules. There can be vessels going directly from the metarterioles to the venules without supplying capillary beds. These vessels form arteriovenous (A-V) shunts and are called arteriovenous capillaries. The thickness of the wall and endothelium of these structures and the proportionate amounts of the various vascular wall components have been shown in Figure 2.7. Capillaries and venules have very thin walls. The capillary, as mentioned before, lacks smooth muscle and has only a layer of endothelium. Smooth muscle and elastic tissue are present in greater amounts in vessels having vasoactive capabilities, such as arterioles. They are also the site of the greatest drop in mean blood pressure. For this reason, arterioles are the principal contributors to peripheral vascular resistance, which can effectively alter cardiac output.

2.4.2 MICROCIRCULATORY FUNCTION

The structural components of the microcirculation are classified into resistance, exchange, shunt, and capacitance vessels. The resistance vessels, comprising the arterioles, metarterioles, and precapillary sphincters, serve primarily to decrease the arterial pressure to the levels of the capillaries to facilitate effective exchange.

Table 2.2   Diameters of Red Blood Cells (RBC) of
             Some Mammalian Species

Species	Body weight (kg)	RBC diameter (μm)
Shrew	0.01	7.5
Mouse	0.20	6.6
Rat	0.50	6.8
Dog	20	7.1
Man	70	7.5
Cattle	300	5.9
Horse	400	5.5
Elephant	2000	9.2

2.4.3   PRESSURE-FLOW RELATIONSHIPS

A bat wing and the mesenteric bed of a small mammal are two popular preparations for studying blood flow in the microcirculation. The flow into the capillaries has
been shown to remain pulsatile or intermittent in nature. It has also been shown that
the rhythmic vasomotor activity of the precapillary sphincters is responsible for the
observed intermittency. The sphincters may also exhibit constriction and dilation in
response to changes in local metabolites, chemicals, or sympathetic stimuli. Together
with the arterioles, the precapillary sphincters serve to adjust the amount of blood
flow to meet the demands in tissues.

Data from Altman and Dittmer (1961) have shown that in more than 100
mammalian species, the red blood cells are of similar size. This is summarized in
Table 2.2.

If we compare the size of red cells from various mammals, we find the perhaps
surprising fact that their diameters seem to be rather uniform and independent of
mammalian body size (Table 2.2).

Additional data on red cell size are found in Altman and Dittmer (1961). Notice
that no single mammal has a red cell diameter over 10 μm. This suggests the
structural size of the capillaries, which are on the same order of magnitude in these
mammals.

CHAPTER 3

Comparative Analysis With Allometry

3.1 MODELING PRINCIPLES AND COMPARISON ACROSS SPECIES

We have seen that there are general similarities in the anatomical structure and the physiological function of the mammalian cardiovascular system. To draw such conclusions based on observations alone is inadequate, in that the physical principles that govern such similarities are not present. To extrapolate one's observation of similarities from one species of a given size to another, one exercises visual "scaling". Thus, scaling can be used to form some kind of criteria to allow extrapolation or interpolation of observations.

Anatomical similarities can be scaled in the simplest fashion by multiplication or division of body length dimensions. Thus, the length of the aorta of a dog L_a is scaled to that of the cat L_a' by a scaling factor a_1,

$$L_a = a_1 L_a' \tag{3.1}$$

Similarly, their heart weights can be scaled by a constant of proportionality a_2,

$$W_h = a_2 W_h' \tag{3.2}$$

Modeling is an important aspect of interpreting observed similarities and prediction of experimental observations *a priori*. To be useful, an experimental test on a model should provide data that can be easily scaled to obtain any existing static or dynamic information on the "prototype". Of course, the pertinent features of the prototype need to be of sufficient accuracy. It is important that the two systems under observation have the same dimensional homogeneity and are within the same regime. For instance, one would compare aortic blood flow and aortic diameter of man to those of the elephant. However, one would not compare blood flow in the aorta of man to blood flow in the femoral artery of the elephant, nor would one compare blood pressure to flow, which is a different dimensional variable. In this case blood flow should be expressed in cubic centimeters per second or milliliters per second and aortic diameter in centimeters.

Comparative anatomical observations of the cardiovascular structures of mammalian species have found considerable geometric similarities. Geometric similarity implies that the shape is the same and that corresponding linear dimensions of the cardiovascular systems are related by a constant scale factor. Thus, the length of the aorta from the aortic valve to the aorta-iliac bifurcation is about 65 cm in a 70-kg man. This length is about 45 cm in a 20-kg dog. The ratio of the two is 65/45 = 1.44. If the diameter of the dog aorta is 1.6 cm, this ratio can be used to calculate the aortic diameter of the man since

$$65/45 = D/1.6 \qquad\qquad (3.3)$$

or D = 2.3 cm, which should be close to the actual size of the human aorta had we measured it. Notice here that the aortic length and diameter are both expressed in centimeters. In addition, both have the body weight exponent of 1/3. One can also scale the total cross-sectional area of the vascular beds, the number of capillaries, or the size of the heart from one mammal to another. Heart weight in mammals is in a constant proportional to body weight. This constant of proportionality gives a scaling factor such that we can obtain the heart weight of any mammal by simply knowing its body weight. As an example, the body weights of man and a rabbit are 70 and 3 kg, respectively; and the man's heart-weight-to-body-weight ratio is 6/1000, or 0.6%, or 420 g, then the heart weight of the rabbit is

$$W_h = 3 \times 0.6\% = 0.018 \text{ kg, or } 18 \text{ g} \qquad\qquad (3.4)$$

If, however, we were to compare 10 different species, such a procedure would be tedious. Allometric equations, as we shall see in the remaining sections of this chapter, provide a solution to this problem.

Apart from geometric similarity, which addresses common anatomic features, one needs also to examine the kinematic similarity. For instance, the heart generates blood pressure and flow waveforms, which are transmitted to the arterial system. These waveforms should possess certain similarity, and their amplitudes should be proportionally scaled. Indeed, as we shall see, not only do aortic blood pressure waveforms in mammals exhibit striking similarity, but they are of the same magnitude (a proportionality constant of unity). Blood flow waveforms are of the same shape, with magnitudes that can be easily scaled.

Thus, for two fluid mechanical systems to function similarly, both geometric and kinematic similarity must be possessed in order for the systems to be dynamically similar. This is where dimensional analysis and the Pi-theorem come into play for the establishment of similarity criteria, as we shall discuss in the next chapter.

3.2 ALLOMETRIC EQUATION AND DEFINITIONS

J. S. Huxley (1932) considered differential growth as the main reason for many of the observed biological transformations. Quantitatively, this resulted in a power law, or the well-known allometric formula

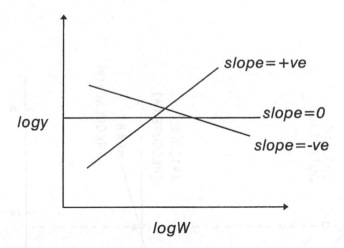

logy

slope=+ve

slope=0

slope=-ve

logW

Figure 3.1 The allometric formula Y = aW^b on a log-log plot with b either positive (variable increasing with body weight) or negative (variable decreasing with body weight).

$$Y = aW^b \tag{3.5}$$

This equation expresses simple allometry. Here Y is any physical variable, a is a constant of multiplication, and b is the exponent. In the special case where the exponent is 0, the physical variable Y is independent of body weight W. When b is 1/3, the variable is said to be dependent on body linear dimensions. When the exponent b is 2/3, the variable is dependent on body surface area. When b = 1, the physical variable is simply proportional to body weight. The inverse proportionality applies when b is negative. On a log-log plot, one can see that a straight line will result with a slope b. The slope can be positive or negative depending on the value of b (see Figure 3.1).

For comparison among species, a physical variable of interest is first measured for each of the species. The values are then plotted on a log-log graph. Since

$$\log Y = \log a + b \log W \tag{3.6}$$

plotting log Y on the ordinate and log W on the abscissa will result in a straight line with slope b and intercept a. As an example of heart rate vs. body weight, Figure 3.2 illustrates this approach.

Allometry is defined as the change of proportions with increase of size both within a single species and between adults of related groups. Huxley's allometric formula relates any measured physical quantity Y to body weight W, with a and b as empirical constants.

The allometric equation proves powerful for characterization of similarities among species. It is effective in relating a physiological phonomenon among mammals of grossly different body weights. A similarity criterion is established when Y, now consisting of either product(s) or ratio(s) of physically measurable variables, remains constant despite changes in body weights, and is dimensionless. This method

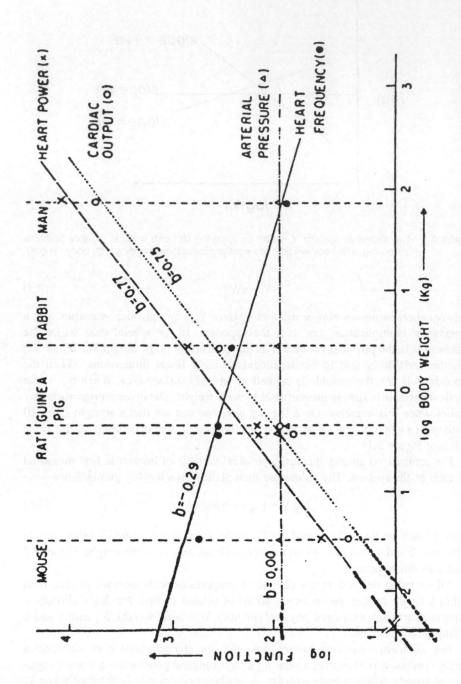

Figure 3.2 Heart rate as a function of body weight on the log-log plot. The slope gives the exponent b, and the intercept gives the constant a.

Table 3.1   Basal Metabolism (cal/day) of Different Animals

	Allometric equation
Mammals	$70.5W^{0.731}$
Birds	$89W^{0.64}$

of establishing similarity criteria is detailed in Chapter 4, Section 4.3. Thus, the exponent b is necessarily zero. In other words, similarity is present whenever any two dimensionally identical measurements occur in a constant ratio to each other. If such a ratio exists among different species, then a similarity criterion is established as the scaling law. Stahl (1963) provided many such criteria based on this approach, as did Gunther (1975) later. For example, if we plot aortic diameter against aortic length for several mammals, the ratio of length to diameter is a constant. This ratio is easily calculated to be about 36.5 from the allometric equations of the aortic diameter,

$$D = 0.48W^{0.34} \tag{3.7}$$

and the length of the aorta (Li, 1987),

$$L_a = 17.5W^{0.31} \tag{3.8}$$

See Figure 3.3.

In general, variations in the empirical values of a and b depend on sample size, the range of body weight differences, and methods of measurement of the physical parameter Y. There have been criticisms on the use of allometry by itself for comparative physiological purposes, on both theoretial (Yates, 1979) and statistical (Smith, 1980; Heusner, 1982) grounds. Indeed, one may obtain dimensionless numbers from allometric ratios with a zero exponent without any physiological significance. For instance, the ratio of aortic length to the length of the index finger is a constant which does not bear any physiological meaning. Therefore, allometry needs to be used in conjunction with physical and physiological principles.

To illustrate allometry, Table 3.1 gives the basal metabolism of different animals in calories per day in relation to body weight expressed in their allometric forms (Brody, 1945). It should be mentioned here that these equations do not indicate how important the energy expenditure is in relation to the structure and function of a particular organ system. Basal metabolism was measured in terms of oxygen and heat production.

Thus, the metabolism of mammals is seen to follow somewhat a 3/4 power law, and larger mammals will have a higher metabolic rate, as expected.

Despite the shortcomings of allometric equations, their use has its fundamental place in comparative physiology. We shall discuss the extent of the usefulness of simple allometry and its limitations when applied to the cardiovascular systems of mammals.

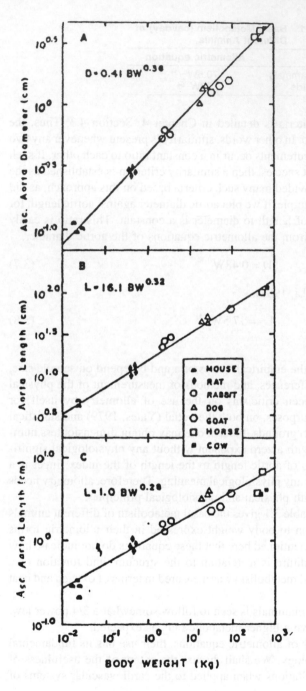

Figure 3.3 Diameter of mammalian aortas plotted against their lengths.

3.3 GROWTH AND DIFFERENTIAL GROWTH

3.3.1 CONSTANT PROPORTIONAL GROWTH

In his work *On Growth and Form*, D'Arcy Thompson (1917) stressed that all organic forms are the result of differential growth. This includes general growth, which may be quantitatively different in space, or localized growth at any particular site. The subject of differential growth received little consideration before Huxley's time. Thompson's treatment, though original and detailed on some subjects and certainly interesting throughout, is nevertheless incomplete. His main original contribution is his use of Cartesian transformations to illustrate the evolution from one form to another. This stimulated many later investigations in biological transformations.

The problem of differential growth, fundamental to biological studies, was intensively investigated by Huxley in the 1920s. He studied certain phases of what he termed "The Problems of Relative Growth" and demonstrated the existence of empirical laws which appear to govern differential growth. He published a book on the subject in 1932.

Huxley first studied a number of known cases of differential growth and attempted to obtain quantitative expressions. He was able to obtain a simple formula, which appears to be a good first approximation of a general law for differential growth. Futhermore, the establishment of one quantitative rule leads to the discovery of others.

Huxley postulated a hypothesis that the ratio between the intensity (or relative rate) of growth of the organ and that of the body remains constant over long periods of the animal's life. Relative growth rate is the rate of growth per unit weight, i.e., the actual absolute growth rate at any instant divided by the actual size at that instant. In other words, there is a constant partition coefficient of growth intensity between organ and body. There are many organs within the mammalian species that grow at rates markedly different from the body as a whole. In these cases, the rate of growth is not distributed uniformly, but may exhibit some regular pattern. Modern pattern recognition techniques, of course, were not available to him at the time. He chose to study the quantitative relation between the magnitudes of the two variables and met with success. To look at this approach from the cardiovascular point of view, we can examine the relation between heart weight and body weight,

$$W_h = aW^b \tag{3.9}$$

If b = 1, then there is proportional growth, and if b differs greatly from 1, then there is obviously a differential growth. The exponent b is found to be very close to 1, suggesting that the heart grows proportionally with increasing body weight.

Concerning growth and differential growth, there are three important findings by Huxley that can be summarized here:

1. Growth is a process of self-multiplication of living substances. In other words, the rate of growth of an organism growing equally in all its parts is at any time proportional to the size of the organism.

2. The rate of self-multiplication slows down with increasing age (size).
3. The rate of self-multiplication is affected by the external environment.

There are some terms that deserve definitions. The formula

$$y = bx^k \tag{3.10}$$

contains the "growth coefficient" k as its exponent. This formula describes the relation of the growth of a part or organ y to that of the whole organism x. An organ that is thus growing at a different rate from the body is said to be "heterogonic" (Pezard, 1918). If it is growing at the same rate as the body, it is "isogonic".

This simple formula appears to be adequate to describe a wide range of growth phenomena, including many cases of changes in proportions during growth in many mammalian species. It also describes the change of adult proportions with increasing absolute size in related species of animals and during the course of evolutionary change, as well as changes in the proportions of various chemical constituents of the growing organism (Huxley, 1932).

3.3.2 DIFFERENTIAL GROWTH AND GROWTH IN TIME

The usefulness of the simple allometry formula, Equation (3.5), is that it does not explicitly contain the time dependence or age. Thus, one can ignore the time relations of growth by relating the sizes of the different parts of the body to the mammal's size, regardless of age, as long as corresponding adult groups are used for comparison. This illustrates a well-known fact that form during growth is a function of absolute size rather than of absolute age, insofar as these two measures of growth are independent. The elimination of time in studying change of form during growth is consistent with the fact that variations in nutrition and growth rate do not in general affect proportions at a given absolute size, and it is reflected in the frequently used concept of physiological age. However, there are many equations for describing growth in time, t, for instance, exponential, linear, or constant rate, respectively,

$$y = e^{at} \tag{3.11}$$

$$y = at \tag{3.12}$$

or

$$dy/dt = a \tag{3.13}$$

where a is a constant.

3.3.3 DEVIATIONS FROM SIMPLE ALLOMETRY

One of the most serious difficulties, and one that is often neglected, in studies of differential growth is how to decide whether the growth trend of the data is adequately represented by a straight line on a log-log or a double logarithmic graph. Although the allometry formula has frequently been accepted as nothing more than a useful first approximation to the true law of differential growth, little attempt has been made to develop a more general formula. Critical points appear as either sudden changes of slope or actual gaps in the allometry line. This is particularly true when interpolation is used. Extrapolation of data points to extremes of body weight differences would also in general induce large errors. For instance, when the ratio of heart weight as a function of body weight measured in large mammals, weighing more than 10 kg, is used to estimate the similar ratio for a very small mammals, weighing less than 100 g, error can be considerable.

Another type of deviation from simple allometry is a rhythmical variation about an average trend of simple allometry, of which several cases were discussed by Huxley. Other problems in the use of simple allometry have been illustrated by Haldane: the different parts of an organ (e.g., segments of a limb) show unequal constant differential growth ratios against body size, then the whole organ cannot exactly obey the allometry law against the same standard, and vice versa. There may be equations that provide a better quantitative description than simple allometry alone. For instance, the DuBois (1916) formula that is widely accepted for the calculation of body surface area,

$$BSA = a_1 W^{b_1} + a_2 H^{b_2} \tag{3.14}$$

which is dependent on body weight W and height H.

This differs greatly from the 2/3 power law,

$$BSA = aW^{2/3} \tag{3.15}$$

3.4 THE IMPORTANCE OF BODY SIZE AND ORGAN SIZE

Now we have seen the general theory behind the laws of growth and differential growth. We see that each organ in its adult form bears a constant relation to body size. This is seen from Table 3.2. We now direct our attention to the mammalian cardio-vascular system. There are some obvious questions: what is the importance of heart size in relation to body size? Why is the aorta a specific diameter? We have seen that heart weight is constantly related to body weight in mammals; what determines the 0.6% proportionality? It appears at first that the organ size is such that its growth is dependent on the physiological functional demand. This aspect can be seen in the growth of the heart from infancy to adulthood. We also know that when the heart is abnormally large, as in the case of cardiac hypertrophy and dilatation, normal cardiovascular function is compromised. This is discussed in Chapter 9.

The cardiovascular system is known to be regulated within limits by functional or mechanical factors and by closely coupled organs, e.g., the heart and the lungs; while further complications occur in the mutual adjustment of the various parts of organs to each other.

Table 3.2   Organ Weight as a Percentage of Body Weight in Mammals

Organ	% of body weight
Heart	0.4–0.6
Brain	1.8–2.0
Lungs	0.8–1.0
Kidney	0.4–0.6

Note: Muscle, skin, and other organs account for more than 90% of body weight.

Table 3.3   Allometric Relations of Brain Size in Some Mammals

Man	$0.08W^{0.66}$
Apes	$0.03W^{0.66}$
Monkeys	$0.02W^{0.66}$
Mammals	$0.01W^{0.70}$

Larger organ size requires higher metabolism. If a larger than normal heart is to eject blood to perfuse the same vascular network, then there will obviously be energy that is "wasted". From this simple point of view, we see that the mammalian cardiovascular system "optimizes" its structure to perform its function. As a consequence, the organ size is optimized to its body weight. On the other hand, a small mammal with a small heart can perform the same amount of work as a large mammal with a large heart. There seems also to exist functional optimality, where ejected blood adequately perfuses its vascular beds in a specified amount of time.

Not all mammalian organs exhibit constant proportions to their body weights. One of the best-known examples is the brain size (Stahl, 1965), as shown in Table 3.3. It appears that, indeed, we *Homo sapiens* have the the largest relative brain size. Different organs also constitute different percentages of body weight. The heart weighs only about a quarter to a third as much as the brain.

3.5 CIRCULATORY ALLOMETRY

Allometric equations have been established by many investigators for various body organs. Among biological systems, the main interest here is in the similarities of the circulatory function of the mammalian species in terms of allometry and hemodynamics. I shall therefore elaborate in some depth the approaches that begin with allometry and conclude with the establishment of similarity principles for the mammalian cardiovascular systems.

From a physiological point of view, there are various laws and equations that govern the structure and function of the heart. It is not an easy task to select precisely the appropriate parameters to describe the function of the heart under a prescribed physiological condition. For the overall function of the heart, the obviously important factors are the heart rate and heart weight, external mechanical work, and metabolic rate. Stroke volume and blood pressure are also of great importance in quantifying

Table 3.4   Some Circulatory Allometric Relations

Y		a	b	Source
Heart rate	f_h	3.60	−0.27	Adolph (1949)
(sec⁻¹)		4.02	−0.25	Stahl (1967)
		2.36	−0.25	Holt et al. (1968)
Metabolic rate	MR	3.39×10^7	0.75	Kleiber (1947)
(erg/sec)		3.41×10^7	0.734	Kleiber (1961)
Stroke volume	V_s	0.74	1.03	Juznic (1964)
(ml)		0.66	1.05	Holt et al. (1968)
Arterial pressure	P	1.17×10^5	0.033	Gunther and Guerra (1955)
(dyn/cm²)				

Table 3.5   Allometric Relations of Cardiovascular
Parameters

Y	a	b
Aortic area (cm²)	1.79×10^{-3}	0.67
Arterial length (cm)	1.78	0.32
Cardiac output (LV) (ml/min)	166	0.79
Cardiac output (RV) (ml/min)	179	0.78
Density (g/cm³)	1.05	0
Duration EKG (P) (sec)	1.39×10^{-2}	0.168
Duration (QRS) (sec)	1.11×10^{-2}	0.165
End systolic volume (LV) (ml)	0.59	0.99
End systolic volume (RV) (ml)	0.62	0.99
End diastolic volume (LV) (ml)	1.76	1.02
End diastolic volume (RV) (ml)	2.02	1.02
Pulse time (sec)	4.3×10^{-2}	0.27
Systemic pressure (dyn/cm²)	1.17×10^5	0.033
TPR (dyn.s.cm⁻⁵)	3.35×10^6	−0.68
Velocity (cm/s)	0.184	0.67
Ventricular weight (LV) (g)	1.65	1.11
Ventricular weight (RV) (g)	0.74	1.06
Viscosity (P)	0.03	0

the function of the heart. There are other variables that can exert a direct influence on the performance of the heart and the vascular system.

Huxley's allometric formula $Y = aW^b$ can be applied to express the parameters of the heart, using body weight as a reference. Some of the parameters that are important to analysis of the performance and energy requirements of the heart are given in Table 3.4. Other examples of allometric equations of some cardiovascular parameters are given in Table 3.5.

Von Hoesslin (1927), and Lambert and Teissier (1927) were probably the first to propose some circulatory similarities. They suggested that individual biological periods such as cardiac cycle ($1/f_h$; f_h = heart rate) are proportional to the length dimensions of the body, i.e., $W^{1/3}$. Deviations from this can be seen in Table 3.4.

Usefulness of allometry in the circulatory system can be easily appreciated. The use of body weight as a reference is a useful concept, particularly in the application of Huxley's (1932) allometric equation. Two examples are shown in Figure 3.4 where the length of the heart cycle and heart rate are plotted against body weight. Some other examples concerning cardiovascular parameters are shown in Figure 3.5. One notices immediately that the smaller the mammal, the smaller the heart weight,

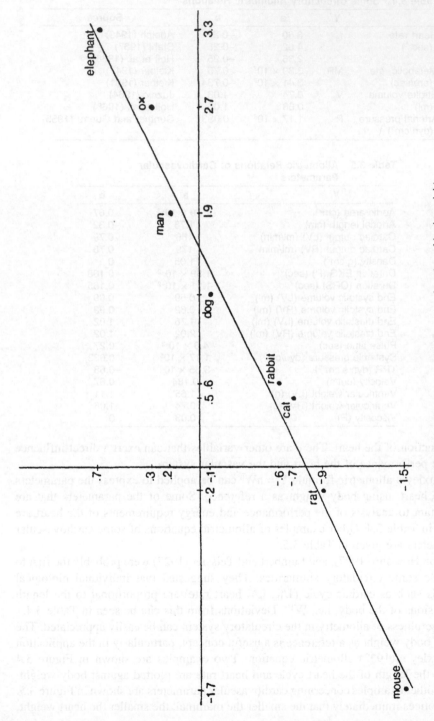

Figure 3.4 Allometric relations showing the measured cardiovascular variables as a function of body weight.

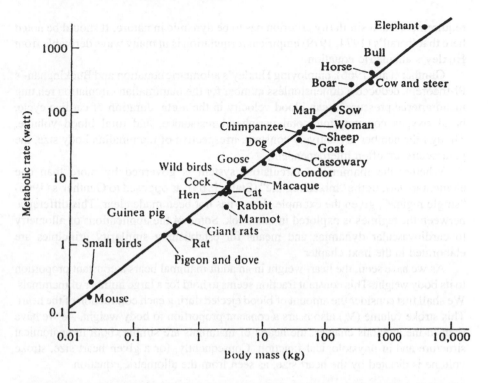

Figure 3.5   Allometric relations showing the measured cardiovascular variables as a function
of body weight.

but the factor the heart rate. Also one should note here that smaller mammals also
have shorter life spans.

The power law equation used by Holt and associates describing left ventricular
function and aortic length are in the same form as Huxley's equation, again using
body weight as a reference. There are many such examples, and readers are referred
to a comprehensive summary by Stahl (1963, 1965).

3.6 INTRODUCTION OF ALLOMETRY TO HEMODYNAMICS

The close resemblance of fluid motion and blood flow has allowed the principles
applied to hydrodynamics be carried over to hemodynamics. Similarity criteria well
established in hydrodynamics were applied to arterial blood flow. For instance, the
Reynolds number is essential for identifying viscous similitude and laminar-to-
turbulent flow transitions (Section 7.1).

Iberall (1974), following Thompson's (1917) proposition that the engineering
design of biological form is a dynamic process, attempted to find dynamic limits in
relating form parameters of mammals to cardiovascular parameters. While the discussions of the empirical formulas found are interesting, the information provided
does not explain pulse wave transmission theory. However, he grasped the important

requirement that a similarity criterion has to be dynamic in nature. It should be noted here that Iberall's (1974, 1979) empirical formulation is in many ways derivable from Huxley's allometric equation.

Gunther et al. (1966), employing Huxley's allometric equation and Buckingham's Pi-theorem, deduced a dimensionless number for the mammalian circulation relating mean arterial pressure, mean blood velocity in the aorta, duration of cardiac cycle, basal oxygen consumption, total peripheral resistance, and total blood volume. Though the number is relatively constant, irrespective of mammalian body size, the parameters are often interdependent.

Whether the mammalian circulatory system is governed by more than one elementary law, or the "mixed regime" (Stahl, 1963), as opposed to Gunther's (1966) "single regime", given the example above, has not been made clear. This difference between the regimes is explored in this book. Some of the applications of allometry to cardiovascular dynamics and means for establishing similarity principles are elaborated in the next chapter.

As we have seen, the heart weight in an adult mammal bears a constant proportion to its body weight. This constant fraction seems to hold for a large number of mammals. We shall first consider the amount of blood ejected during each contraction of the heart. This stroke volume (V_s) also bears a constant proportion to body weight. As we have seen in the previous chapter, the hearts of mammals are similar, both in anatomical structure and in physiological function. Consequently, for a given heart size, stroke volume is dictated by the heart size, as seen from the allometric equation

$$W_h(LV) = 2.61W^{1.10} \text{ g} \tag{3.16}$$

$$V_s = 0.66W^{1.05} \text{ ml} \tag{3.17}$$

or

$$V_s = 0.74W^{1.03} \text{ ml}$$

Hemodynamics in the broadest sense is the study of the dynamics of blood flow. The total amount of blood flow out of the ventricle during ejection is therefore equal to the stroke volume. Mathematically speaking, the stroke volume is the integral of flow during the ejection period,

$$V_s = \int Q(t)\,dt \tag{3.18}$$

where $Q(t)$ is the instantaneous blood flow measured in the ascending aorta. Equivalently, this flow is the rate of change of ventricular volume,

$$Q(t) = dV(t)/dt \tag{3.19}$$

Stroke volume has been considered one of the most important hemodynamic quantities in assessing ventricular function. Together with blood pressure, its magnitude bears a direct relation to the energy expediture of the heart. The external work (EW), or the work performed by the heart to overcome any load during ejection, is given by

$$EW = PV_s \qquad (3.20)$$

Blood pressures in mammalian aortas are invariant of body weights. In allometric form, the mean arterial pressure is expressed as

$$p = 1.17 \times 10^5 \, W^{0.033} \, \text{dyn/cm}^2 \qquad (3.21)$$

or

$$p = 87.8 W^{0.033} \, \text{mmHg}$$

The left ventricular EW can then be easily computed from the product of the mean arterial pressure and stroke volume,

$$EW = 0.87 \times 10^5 \, W^{1.063} \, \text{erg}$$
$$= 0.0087 \, W^{1.03} \, \text{J} \qquad (3.22)$$

noting that $1 \, J = 10^7$ erg. Therefore, a larger ventricle will generate a greater amount of external work, simply because of its larger heart size.

As we noted earlier, Harvey in 1626 wrote in his now-famous book *De Motu Cordis* that the heart contracts during one third of the cardiac period, known as systole, and rests to allow filling during the remaining two thirds of the cardiac period, the diastole. Thus, the contraction of the heart leads to the pulsatile nature of pressure and flow. Pressure pulse generated in systole and in diastole comprises the pulsation, and their maximal differential is the pulse pressure. This is shown in Figure 3.6. There has been a considerable amount of work relating the pulse pressure waveform to blood flow. Consequently, aortic pressure waveform has been used to obtain the stroke volume. Technological advances have made the thermodilution measurement of stroke volume a common practice. Electromagnetic and ultrasound Doppler transducers are also common instruments used to obtain flow. Arterial hemodynamics, which we shall examine later, also deals with pressure-flow relationships and pulse transmission characteristics in vascular beds.

Cardiac output (CO), a familiar parameter used in physiology and medicine to assess cardiac function, is defined as the amount of blood ejected per minute, and equivalently,

$$CO = V_s f_h \qquad (3.23)$$

or

$$CO = (0.74 W^{1.03})(4.02 W^{-0.25}) 60/1000$$
$$= 0.178 W^{0.78} \, \text{l/min} \qquad (3.24)$$

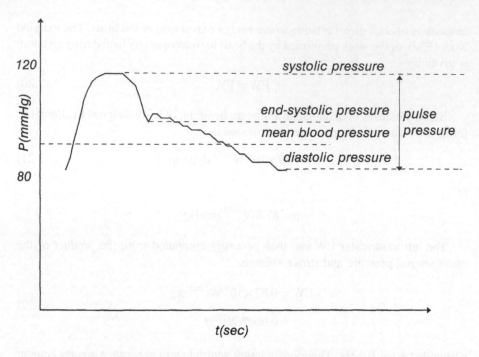

Figure 3.6   Pressure waveform in the aorta showing the definitions of systolic blood pressure, diastolic blood pressure, and their difference; the pulse pressure; and the mean pressure, average value over the entire cardiac cycle.

since

$$f_h = 4.02 W^{-0.25} \, s^{-1} \tag{3.25}$$

For a 70-kg human (W = 70), Equation (3.24) gives a value of about 5 l/min, a very good estimation. For a 20-kg dog, this value is 1.85 l/min, and for a 2000-kg elephant, it is some 67 l/min. Cardiac output calculated on the basis of this formula is shown in Table 3.6.

The total peripheral resistance for a mammalian systemic arterial tree can be calculated from the relation

$$R_s = p/CO = 2.8 \times 10^6 \, W^{-0.747} \, \text{dyn s cm}^{-5} \tag{3.26}$$

Thus, the peripheral resistance follows the $-3/4$ power and is inversely proportional to the metabolic rate $(+3/4)$.

These examples show that simple manipulations of allometric formulae, such as that demonstrated here, can produce a new and useful set of allometric equations. In the present example, we can easily obtain a good estimate of the peripheral resistance of the systemic circulation of a mammalian species, as long as the body weight is known. This is far easier than carrying out measurements in order to obtain this value

Table 3.6 Cardiac Output of Some Mammalian Hearts Based
on the Allometric Equation CO = 0.178W$^{0.78}$ l/min

Species	Body weight (kg)	Cardiac output (l/min)
Elephant	2000	67
Horse	400	19
Man	70	5
Dog	20	1.8
Rabbit	3.5	0.5
Mouse	0.25	0.06
Tree shrew	0.005	0.003

(an impossible task in the tiny tree shrew). This derived allometric equation compares favorably to those reported by, for instance, Gunther and Guerra (1955), who gave

$$R_s = 3.35 \times 10^6 \, W^{-0.68} \, \text{dyn s cm}^{-5} \qquad (3.27)$$

The scope of hemodynamics, which we shall cover in subsequent chapters, places greater emphasis on the heart and the arterial system, including such topics as the contractility of the heart and arterial pulse wave transmission.

CHAPTER 4

Dimensional Analysis For Identifying Circulatory Similarities

4.1 BASIC MATHEMATICAL TOOLS

I have made an effort to include the mathematical methods that are applicable to our interest in establishing similar criteria and performing comparative analysis. The treatment is aimed at fundamental concepts and at providing a foundation for problem solving. Readers are encouraged to consult other books on related topics for more detailed coverage.

4.1.1 EXPONENTIAL AND LOGARITHMIC FUNCTIONS

The allometric equation

$$y = aW^b \tag{4.1}$$

has an "exponent" b, and the value of the variable y is dependent on the numerical values of both the constant a and the exponent b. In many biological applications, the word *exponent* is often used instead of *index*. I shall use *exponent* in this book leaving *index* to the numerical representation of a physiological phenomenon or state, such as the *cardiac index* or the *index of contractility*.

Functions in which the variable is in the exponent, for instance,

$$10^x, e^{-t/T}, e^{j(wt+\theta)} \tag{4.2}$$

are called exponential functions. Many of the exponential functions are those of spatial and temporal descriptions of physiological processes. These are commonly expressed in terms of time (t) and distance (x). The exponent, as examples illustrated above, can be either real or complex (with $j = \sqrt{-1}$), positive or negative. Many biological phenomena can be described by exponential functions that are frequently encountered in nature. The simple allometric equation (4.1) with an exponent value of b is also known as the power law formula (W to the power b). This relation has a curve as shown on a linear-linear or lin-lin plot (Figure 4.1).

0-8493-0169-6/96/$0.00+$.50
© 1996 by CRC Press, Inc.

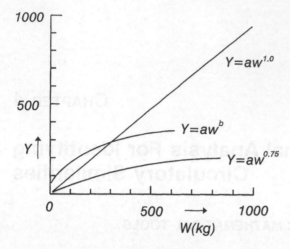

Figure 4.1 The formula $y = aW^b$ on a linear-linear plot.

Given two exponential functions

$$y_1 = a_1 W_1^{b_1} \quad \text{and} \quad y_2 = a_2 W_2^{b_2} \tag{4.3}$$

addition or subtraction can be directly performed if $W_1 = W_2 = W$ and $b_1 = b_2 = b$, i.e.,

$$y_1 + y_2 = a_1 W^b + a_2 W^b = (a_1 + a_2)W^b \tag{4.4}$$

Multiplication or division can be directly performed if $W_1 = W_2 = W$;

$$y_1 y_2 = a_1 a_2 W^{(b_1+b_2)} \tag{4.5}$$

$$y_1/y_2 = (a_1/a_2)W^{(b_1-b_2)} \tag{4.6}$$

Now we can look at logarithmic functions. For an exponential function

$$N = x^k \tag{4.7}$$

then

$$\log_e N = \log_e x^k = k \log_e x \tag{4.8}$$

This is taking the logarithm of the number N to the base e. The other common base one can also take is base 10. The following abbreviations have become standard practice:

$$\log_e = \ln \tag{4.9}$$

$$\log_{10} = \log \tag{4.10}$$

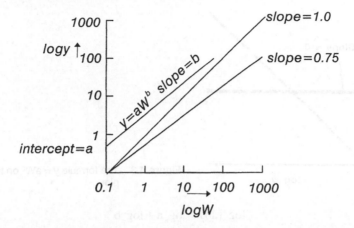

Figure 4.2   The formula y = aW^b on a log-log plot showing the linear relation with slope b and intercept a.

Power law or exponent function is easily converted to logarithmic functions. For

$$y = aW^b$$

taking logarithm to the base 10 of both sides give

$$\log y = \log a + \log W^b$$

or

$$\log y = \log a + b \log W \qquad (4.11)$$

This approach gives the allometric relation a linear line on a logarithmic-logarithmic or log-log graph (Figure 4.2).

A special condition exist when b = 0, i.e.

$$y = aW^0 \qquad (4.12)$$

or when the variable is independent of mammalian body weight,

$$\log y = \log a + 0 \log W$$
$$= \log a \qquad (4.13)$$

or on the log-log plot, one gets a horizontal line with a constant value a (Figure 4.3). The following properties become apparent,

$$\log_{10}(ab) = \log_{10} a + \log_{10} b$$
$$= \log a + \log b \qquad (4.14)$$

Similarly, to the base e,

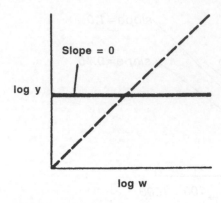

Figure 4.3   The formula y = aW⁰ on the log-log plot.

$$\log_e(ab) = \log_e a + \log_e b$$
$$\ln(ab) = \ln a + \ln b \tag{4.15}$$

Also, for division, we have

$$\log(a/b) = \log a - \log b \tag{4.16}$$

Another useful formula is

$$\log_a b = \log_c b / \log_c a \tag{4.17}$$

which allows the change of base from a to c. It follows that

$$\log_e x = \log_{10} x / \log_{10} e \tag{4.18}$$

A common conversion is in the expression of allometric equations using body weight W in grams (g) instead of kilograms (kg). These bases can be easily interchanged. Suppose W is in grams and we would like to change W to kilograms:

$$y = aW^b \ (\text{W in g})$$
$$y = \left(a \times 1000^b\right)\left(W^b / 1000^b\right)(\text{W in kg}) \tag{4.19}$$

Notice that everything stays the same except that a factor of 1000^b is involved in the conversion. Similarly, if one wishes to use W in kilograms instead of grams, we have

$$y = aW^b \ (\text{W in g})$$
$$y = \left(a/1000^b\right)\left(1000^b W^b\right)(\text{W in kg}) \tag{4.20}$$

Now some examples of the use of exponential and logarithmic functions.

$$\log_{10} 10^x = x \tag{4.21}$$

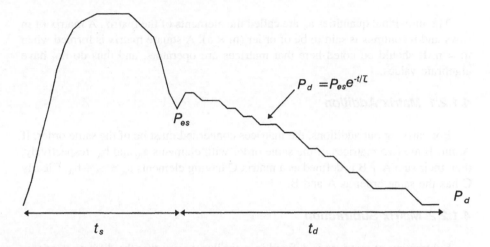

Figure 4.4 Aortic pressure showing monoexponential decay with a time constant τ.

$$\log_e e^{-t/\tau} = -t/\tau \tag{4.22}$$

To calculate the time constant τ for diastolic aortic pressure decay from end-systolic pressure (p_{es}) at 108 mmHg to end-diastolic pressure (p_d) of 80 mmHg during the diastolic period t_d of 0.6 s (Figure 4.4), we have

$$p_d = p_{es} e^{-td/\tau} \tag{4.23}$$

Rearrange and take logarithm of both sides,

$$\ln(p_d/p_{es}) = \ln(e^{-td/\tau}) \tag{4.24}$$
$$= -t_d/\tau$$
$$-0.3 = -0.6/T$$

or

$$\tau = 2\,s \tag{4.25}$$

4.1.2 MATRICES

A matrix is a set of mn quantities arranged in a rectangular array of m rows and n columns. The matrix is denoted by a single letter, say, A, such that

$$A = \begin{pmatrix} a_{11} & a_{12} & \cdots & a_{1n} \\ a_{21} & a_{22} & \cdots & a_{2n} \\ \vdots & \vdots & & \vdots \\ a_{m1} & a_{m2} & \cdots & a_{mn} \end{pmatrix} \tag{4.26}$$

The individual quantities a_{ik} are called the elements of the matrix. A matrix of m rows and n columns is said to be of order (m × n). A square matrix is formed when m = n. It should be noted here that matrices are operators, and thus do not have algebraic values.

4.1.2.1 Matrix Addition

For carrying out additions, the matrices concerned must be of the same order. If A and B are two matrices of the same order with elements a_{ik} and b_{ik}, respectively, then their sum A + B is defined as a matrix C having elements $c_{ik} = a_{ik} + b_{ik}$. Clearly C has the same order as A and B.

4.1.2.2 Matrix Subtraction

Subtraction of matrices is defined in a similar way in that the difference of two matrices A and B of the same order is a matrix D whose elements $d_{ik} = a_{ik} - b_{ik}$.

The elements of matrices can be real or complex numbers. It is clear that the laws of addition and subtraction of elementary algebra also apply to matrices since their sum and difference are defined directly in terms of the addition and subtraction of their elements. In mathematical terms, addition and subtraction of matrices are both associative and commutative.

4.1.2.3 Matrix Multiplication

The requirement of matrix multiplication is such that two matrices A and B can only be multiplied together to form their product AB when the number of columns of A is equal to the number of rows of B. These matrices are called conformable matrices. Suppose A is a matrix of order (m × q) with elements a_{ik} and B is a matrix of order (q × n) with elements b_{ik}. Then their product AB is a matrix C of order (m × n) with elements c_{ik}. For example, if A and B are (3 × 2) and (2 × 2) matrices, respectively, then their product, C = AB, is a (3 × 2) matrix. Matrices satisfy the associative and distributive laws of multiplication, provided that the products are defined.

4.1.2.4 Unit Matrices

The unit matrix is a diagonal matrix with all its diagonal elements equal to unity. It is usual to denote such matrices by the letter I; for example, the (3 × 3) unit matrix is

$$I = \begin{pmatrix} 1 & 0 & 0 \\ 0 & 1 & 0 \\ 0 & 0 & 1 \end{pmatrix} \tag{4.27}$$

It is obvious that multiplying any matrix by a unit matrix leaves the matrix unchanged provided the product is defined.

4.1.2.5 The Adjoint and Inverse Matrices

If A is the square matrix, then its adjoint (usually denoted by adj A) is defined as the transpose of the matrix of its cofactors. In other words, if A_{rs} is the cofactor of the element a_{rs} (i.e., the value of the determinant formed by deleting the row and column in which a_{rs} occurs and attaching a plus or minus sign), then the matrix of the cofactors is the square matrix B where

$$B = \begin{pmatrix} A_{11} & A_{12} & \cdots & A_{1n} \\ \vdots & \vdots & & \vdots \\ A_{n1} & A_{n2} & \cdots & A_{nn} \end{pmatrix} \quad (4.28)$$

If follows that the matrix A^{-1} defined by

$$A^{-1} = \text{adj} A / |A| \quad (4.29)$$

has the property that

$$AA^{-1} = I \quad (4.30)$$

and is referred to as the inverse matrix of A. It is easily verified that multiplication of matrices and their inverses is commutative in that

$$AA^{-1} = A^{-1}A = I \quad (4.31)$$

Clearly only nonsingular square matrices have reciprocals, since A^{-1} is undefined when $|A| = 0$.

$$|C| = |(A - \lambda_0 I)| = 0 \quad (4.32)$$

is the characteristic equation of the matrix. The roots of this equation or λ_0's are known as eigen values.

4.1.2.6 Determinants

The determinant of a matrix A is defined only when A is square and is usually denoted by $|A|$ or det A. For example, if

$$A = \begin{pmatrix} 1 & 0 & 1 \\ 2 & 3 & 1 \\ 3 & 1 & 2 \end{pmatrix} \quad (4.33)$$

then

$$A = \begin{vmatrix} 1 & 0 & 1 \\ 2 & 3 & 1 \\ 3 & 1 & 2 \end{vmatrix} = -2 \tag{4.34}$$

The determinant of a unit matrix is therefore 1.

4.1.3 SOLUTION OF LINEAR EQUATIONS

We will use quite extensively, the techniques of solving for several unknown variables in simultaneous equations. We shall elaborate the logical approaches to arrive at the solutions in later chapters, but will introduce the mathematical concept here. Consider the set of n linear equations in the n unknowns x_1, x_2, ..., x_n

$$a_{11}x_1 + a_{12}x_2 + \cdots + a_{1n}x_n = k_1$$

$$a_{21}x_1 + a_{22}x_2 + \cdots + a_{2n}x_n = k_2 \tag{4.35}$$

$$a_{n1}x_n + a_{n2}x_2 + \cdots + a_{nn}x_n = k_n$$

where the coefficients a_{ik} and the k_i are known constants. Using

$$AX = K \tag{4.36}$$

concept of matrix multiplication defined earlier it is clear that these equations can be written in matrix form as

$$\begin{pmatrix} a_{11} & a_{12} & \cdots & a_{1n} \\ a_{21} & a_{22} & \cdots & a_{2n} \\ \vdots & \vdots & & \vdots \\ a_{n1} & a_{n2} & \cdots & a_{nn} \end{pmatrix} \begin{pmatrix} x_1 \\ x_2 \\ \vdots \\ x_n \end{pmatrix} = \begin{pmatrix} k_1 \\ k_2 \\ \vdots \\ k_n \end{pmatrix} \tag{4.37}$$

or

$$AX = K \tag{4.38}$$

$$IX = XA^{-1}K \tag{4.39}$$

Equation (4.39) is therefore the solution of the matrix Equation (4.38) and enables the values of x_1, x_2, ..., x_n to be found directly by evaluating $A^{-1}K$.

There are many approaches to arriving at the solutions of linear differential equations with several unknowns. Our interest here is to use the techniques to obtain the dimensionless Pi numbers to establish similarity criteria. Illustrations of step-by-step solutions are presented in Chapter 7.

4.1.4   FOURIER ANALYSIS

Many biological signals and processes are periodic in nature. The cardiovascular system operates in a rhythmic fashion. The beat-to-beat contraction of the heart and the pulsatile pressure and flow pulse transmission are integral parts of the rhythm. Jean Baptiste Fourier (1882) introduced this concept over a century ago of decomposing time domain periodic signals into their components, known as the Fourier series. Any periodical signal that satisfies the convergence conditions given by Dirichlet can be expressed in its Fourier series. Fourier analysis is useful in that it permits us to look at the frequency content of the signals and allows a greater in-depth analysis of the operating characteristics of the cardiovascular system.

I shall provide the fundamental approaches for the analysis of cardiovascular signals. To use the Fourier series, two basic requirements need to be met: periodicity and linearity. For cardiovascular dynamic signals, periodicity is satisfied. Under normal physiological conditions, the cardiac period stays about constant from beat to beat. Attinger et al. (1966), for instance, found that the heart period and respiratory cycle vary by 1 to 3%. This introduces a maximum spurious harmonic content of less than 6%. There is no significant difference in the magnitude or phase of the Fourier coefficients whether one chose to analyze over one cardiac cycle or over the whole respiratory period. The linearity requirement is satified when the blood flow velocity (v) is small compared to the pulse wave velocity (c) or the radial distension (dD) of the blood vessel is small compared to the diameter (D) of the blood vessel. This latter is referred to as the "small strain" condition.

$$v/c \ll 1$$
$$dD/D \ll 1$$

(4.40)

I shall provide examples of the use of Fourier analysis that are relevant to our interest. Consider aortic blood pressure waveform in Figure 4.4. Pressure at any point in one cardiac cycle is repeated at the same point at an integral cardiac period kT apart; k is an integer, positive or negative. One can express this mathematically as

$$p(t) = p(t + kT) \qquad (4.41)$$

where T is the cardiac period. Expressing in terms of the Fourier series, this becomes

$$p(t) = \bar{p} + \sum_{n=1}^{N} \left(a_n \cos n\omega t + b_n \sin n\omega t \right) \qquad n = 1, 2, \dots N \qquad (4.42)$$

where $\omega = 2\pi/T = 2\pi f_h$. N is the total number of terms. $\bar{p}$ is the mean pressure.

Thus, blood pressure waveform is composed of n Fourier components. These are referred to as harmonics. For n = 1, it is called the first or the fundamental harmonic with a frequency of f Hz; n = 2, the second harmonic with a frequency of $2f_h$; n = 3,

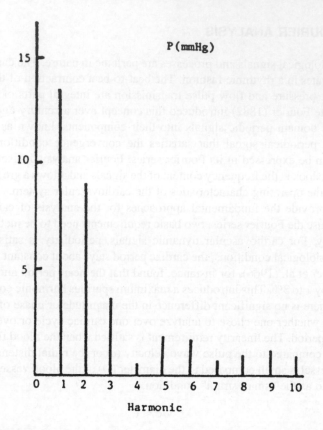

Figure 4.5　Harmonic magnitudes of the aortic pressure waveform represented in the linear spectrum.

the third harmonic with a frequency of $3f_h$, ..., etc. Each harmonic component is associated with a magnitude

$$P_n^2 = a_n^2 + b_n^2 \tag{4.43}$$

and phase

$$\theta_n = \tan^{-1}(b_n/a_n) \tag{4.44}$$

or simply

$$p(t) = \bar{P} + \sum_{n=1} P_n \cos(n\omega t - \theta n) \tag{4.45}$$

Another way to express this series is to display the magnitude and phase in a linear spectrum. An example of the magnitude of the harmonic components of a decomposed aortic pressure waveform is shown in Figure 4.5.

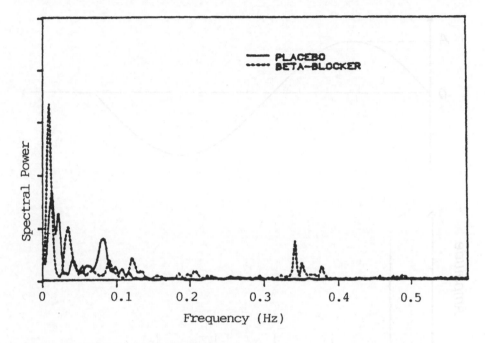

Figure 4.6   Heart rate variability (HRV) spectrum.

It is clear that Fourier analysis is used to transform time domain series into frequency domain information. An example of the spectral analysis of heart rate variability is shown in Figure 4.6. When the cardiac period or R-R intervals are constant, one would only see one harmonic component at the heart rate frequency. Notice there is a peak that appears at a much higher frequency, which corresponds to the respiratory activity.

A pure sinusoidal or a cosinusoidal wave has only one frequency. Thus, its spectrum has only the fundamental (n = 1) frequency component, belonging to the amplitude of the sine or cosine wave (Figure 4.7).

4.2 DIMENSIONAL ANALYSIS AND THE PI-THEOREM OF BUCKINGHAM

Dimensional analysis has its well-founded place in the physical sciences. However, it has scarcely been applied to comparative physiological studies. This is mainly because of the quantitative or analytical nature of problem solving and the earlier lack of interest at an interdisciplinary level. Methods in deriving similarity criteria have come by way of Huxley's allometric equation. Thus, when the exponent b equals 0, a scaling factor is obtained. This scaling factor, however, is not necessarily dimensionless.

Around the turn of the century, Buckingham (1914) proposed some similarity criteria rules, as did Lord Rayleigh at about the same time. In 1915 Buckingham

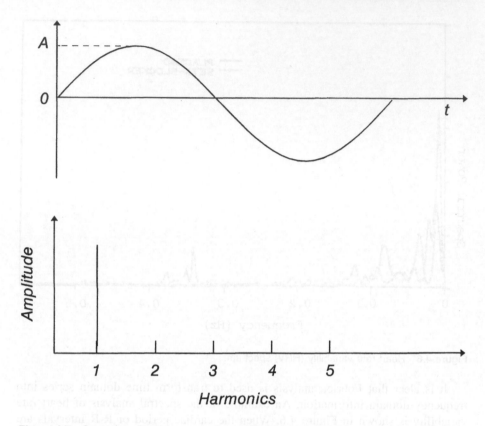

Figure 4.7 Spectrum of a sine wave with amplitude A and phase 0.

published an article to provide a quantitative approach to obtain the so-called Pi-numbers, which are, by definition, dimensionless. Rayleigh indices are extensively used in the fundamental Pi-theorem of dimensional analysis. The theorem states that if a physical system can be properly described by a certain set of dimensional variables, it may also be described by a lesser number of dimensionless parameters which incorporate all the variables. The theorem was later examined by investigators, but its proof is no simple matter.

Use of the Pi-theorem requires all physical quantities be expressed in M (mass), L (length), and T (time), the so-called MLT system. Here we must differentiate between physical quantities and physical constants. The former always possess units, while the latter are not always dimensionless (e.g., Planck's constant). The theorem has wide applications, as will be shown later.

Dimensional homogeneity is another requirement to use the Pi-theorem proposed earlier by Fourier (1882), who stated that any equation applied to physical phenomena or involving physical measurements must be dimensionally homogeneous. Its usefulness can be found in the Navier-Stokes equations describing incompressible fluid flow in the longitudinal direction, for instance (in cylindrical coordinates).

Table 4.1 Dimensional Composition of Some Biological Variables, Listed Alphabetically and Expressed in the MLT system

Physical quantity	Symbol	Dimension		
		M	L	T
Area	A	0	2	0
Diameter	D	0	1	0
Density	ρ	1	-3	0
Elasticity	E	1	0	-2
Flow	Q	0	3	-1
Frequency	f	0	0	-1
Length	L	0	1	0
Metabolic rate	MR	1	-2	-3
Pressure	p	1	-1	-2
Radius	r	0	1	0
Stress	σ	1	0	-2
Stroke volume	V_s	0	3	0
Velocity	v	0	1	-1
Velocity of sound	c	0	1	-1
Viscosity	η	1	-1	-1
Volume	V	0	3	0
Wall thickness	h	0	1	0
Work	W	1	1	-2

Every term in the equation has the dimension of a pressure gradient, or $ML^{-2}T^{-2}$ (see Table 4.1).

Many dimensionless numbers have found their way through the use of the dimensional matrix. The matrix comprises columns representing physical quantities, while rows are filled with basic units (M, L, T). To form a dimensional matrix, *a priori* knowledge of pertinent physical parameters is necessary. If there are eight physical quantities that are important for the description of blood flow in arteries, and there are three basic units (M, L, T), we would be able to obtain $8 - 3 = 5$ dimensionless numbers. In general, the number of dimensionless Pi-numbers is determined by the number of physical quantities minus the rank of the dimensional matrix.

4.2.1 THE [M] [L] [T] SYSTEM

To use the MLT system, one needs to first express explicitly any variable in its physical unit, either using the CGS (cm, g, s) or the MKS (m, kg, s) system or SI units of representation. For instance, blood pressure is commonly measured in millimeters of mercury and must be converted to $g/cm/s^2$. Thus, pressure (p) is given as force per unit area,

$$p = F/A = [M][L]^{-1}[T]^{-2} \tag{4.46}$$

where A has the dimension of cm^2, or $[L]^2$, and force is mass times acceleration,

$$F = ma = [M][L][T]^{-2} \tag{4.47}$$

where a is the acceleration in $cm/s/s = cm/s^2$, or $[L][T]^{-2}$.

The left ventricular volume V is in milliliters or cubic centimeters, and has a dimension of

$$V = [L]^3 \tag{4.48}$$

the aortic flow Q is in milliliters per second or

$$Q = [L]^3 / [T] \tag{4.49}$$

Linear flow velocity has the dimension of

$$v = [L]/[T] \tag{4.50}$$

in centimeters per second.

Heart rate in beats per minute or second has the dimension of

$$f_h = [T]^{-1} \tag{4.51}$$

Table 4.1 gives a list of variables that are expressed in the MLT system.

4.2.2 DIMENSIONAL MATRIX

When formulating a dimensional matrix, it is necessary to identify the parameters that are considered pertinent to the problem at hand. These parameters need to to be expressed in terms of [M], [L], and [T]. For example, given arterial blood pressure (P), flow (Q), and heart rate (f_h), a dimensional matrix can be formed:

$$
\begin{array}{c c c c}
 & P & Q & f_h \\
M & 1 & 0 & 0 \\
L & -1 & 3 & 0 \\
T & -2 & -1 & -1 \\
\end{array}
\tag{4.52}
$$

This is therefore, a 3×3 matrix, or a square matrix.

As another example, suppose that one wishes to examine the relationship between wall tension (T) and the measurable parameters of left ventricular diameter or radius (r) and left ventricular pressure (knowingly, this is the Laplace's law), then a dimensional matrix can be formed in terms of these three parameters, prior to the application of Buckingham's Pi-theorem. This dimensional matrix is as follows:

$$
\begin{array}{c c c c}
 & T & P & R \\
M & 1 & 1 & 0 \\
L & 0 & -1 & 1 \\
T & -2 & -2 & 0 \\
\end{array}
\tag{4.53}
$$

Again, this is a 3×3 square matrix.

The Pi-theorem is not without its limitations. It requires defined pertinent physical parameters or quantities to be known first, yet the derived dimensionless numbers are not at all predictable and are frequently not invariants of the system under consideration (Li, 1983; Rosen, 1983). Although dimensionless, the derived Pi-numbers are not necessarily independent of body weights.

4.3 METHODS FOR ESTABLISHING SIMILARITY PRINCIPLES

Dimensionless numbers provide useful scaling laws, particularly in modeling (such as similarity transformation). Dimensional analysis is a powerful tool, not limited to just mathematics, physics, and modeling, but has potential applicability to biological phenomena.

Despite its many possible applications, dimensional analysis is not without shortfalls. For a given set of physical quantities and basic units, we can generate new dimensionless numbers, which are not necessarily always invariant for a given system. They cannot, therefore, be regarded as similarity criteria. The definition of dimensionless numbers as similarity criteria (Stahl, 1963) is therefore inadequate. Any similarity criteria would need to be both an invariant and dimensionless.

To apply a dimensional matrix to obtain invariant dimensionless numbers, one would have to know beforehand the physical quantities of importance. Thus, one cannot predict the outcome, nor can one know whether the chosen quantities are of prime importance or unique. This is a serious limitation of dimensional analysis.

It has become apparent that living beings are governed by physical laws. There is little doubt that the function of a biological system complements its structure. More than a century ago, von Hoesslin (1888) already attempted to apply the principles of mechanical similarity to living beings, and he spent the next 50 years researching biological systems in related organisms. He concluded that biological systems do function in similar manners.

D'Arcy Thompson's (1917) work on "growth and form" discussed extensively the structural similarities, transformation laws, and scaling of biological organisms. The approach, though not mathematically vigorous, nevertheless made a giant step to modern biological similarity analysis. Huxley's (1932) treatise on "problems of relative growth" was no doubt influenced by Thompson's work. His use of allometric equations and discussions on differential growth patterns are of fundamental importance, as we have seen in Chapter 3. Their work led to later studies by P. B. Medowar, E. F. Adolph, A. V. Hill, and B. Gunther and E. Guerra. The work of von Hoesslin (1927) on mechanical similarities and that of Lambert and Teisser's (1927) on electromechanical similarities have also contributed significantly to modern biological similarity studies.

Stahl (1963) has provided a comprehensive historical review and discussion of various approaches toward establishing biological similarities. But the efforts of earlier British investigators have undoubtedly paved the thoroughfare to more recent research. One example of obtaining similarities was given by Gunther and Guerra (1955), who deduced some "biological constants" by comparing two functions having the same exponent b in the allometric form. The approach is interesting, but they did not provide any explanation of the functional significance of these constants. For instance, they showed that the duration of one respiratory cycle

$$f_r = 44.67W^{-0.29} \tag{4.54}$$

divided into the duration of one cardiac cycle

$$f_h = 234.4W^{-0.268} \tag{4.55}$$

results in a constant ratio of about 5. This indicates that the heart beats five times for each respiration. However, there was no explanation given as to the functional implication that the ratio needs to be 5.

Recent comparative physiological studies have led to postulates that function-related structural design characteristics are governed by similarity laws across mammalian species. Methods leading to the establishment of these laws, however, have been limited. Studies of functional similarities of the mammalian cardiovascular system have not been extensive. Two avenues lead to the establishment of similarity criteria. One is the pursuit of classical allometry and dimensional analysis. The other is the consideration of the dynamics of the cardiovascular system. It is clear that the combination of the two should present a more powerful and a more comprehensive approach. Experimental evaluations of the established similarity criteria of underlying physiological control mechanisms are also lacking. In addition, optimal design features of the mammalian cardiovascular system that correspond to functional dynamics have not been fully identified.

Despite these shortcomings, the combination of allometry and dimensional analysis appears to be a promising avenue leading to the establishment of new similarity criteria as demonstrated by Gunther and De La Barra (1966), Young and Cholvin (1966) and Li (1983, 1986, 1988).

A similarity principle should be both dimensionless and invariant of mammalian body weights. This is to provide a scaling factor. Scaling and biological similarities have been discussed in great detail by Stahl (1963, 1963a) and Gunther (1972, 1975), and recently by Calder (1981) and Schmidt-Nielsen (1986). The establishment of a similarity principle and its utilization support Thompson's (1917) original hypothesis that "closely related organisms should be similar" (also Rosen, 1978). However, there are few such principles derived for the mammalian circulatory system.

Similarity principles that are established should have the following characteristics: (1) the principles should govern certain specified mammalian cardiovascular functions, (2) they should be able to explain the underlying physiological processes or mechanisms that influence the parameters contained in the similarity criteria, (3) they should be applicable to selected species of mammals, and (4) they should be able to explain the optimal functional design features of the mammalian circulation.

4.4 ILLUSTRATIVE EXAMPLES

Circulatory similarities have been derived from various approaches. They have been borrowed mostly from well-known hydrodynamic principles. The use of fluid mechanic similarity criteria, though useful, is too restricted in defining hemodynamic similarities. It is generally limited to local flow regime, and often fails to explain modern arterial pulse wave transmission. A review of pulse transmission characteristics are given by Li et al. (1981) and Li (1987).

In studying blood flow, any two flow fields are similar only if they possess geometric, kinematic, and dynamic similitudes. In this state, the ratios of velocities and forces are constant, and the geometry of the motion of one is merely a constant scale of the other, throughout corresponding points of the entire motion. These similarities, however, have not been extensively examined for the mammalian circulation, except by Li (1988).

It is evident that, even without going into the mathematics of Buckingham's Pi-theorem, one can see that Laplace's law emerges as

$$T = Pr \tag{4.56}$$

The corresponding Pi-number is

$$\pi_1 = T/Pr \tag{4.57}$$

which is dimensionless.

CHAPTER 5

Cardiac Mechanics

5.1 CARDIAC MUSCLE MECHANICS

5.1.1 SARCOMERES AND THE LENGTH-TENSION RELATIONSHIP

Sarcomeres are the fundamental building blocks of the myocardium, as we have learned in Chapter 2. Sarcomeres among mammals exhibit structural and functional similarities. Many of the original studies on the dynamic properties of the sarcomeres, however, were performed on the frog, a nonmammalian species. It is interesting to note that the relation between sarcomere length and isometric tension development in single frog skeletal muscle fibers was developed by yet another Huxley and his associates (Gordon et al., 1966). H. E. Huxley (1958), who wrote on the contraction of muscle, is closely related to J. S. Huxley (1932), who wrote on allometric growth.

It is now known that within the sarcomere, there are thick and thin filaments. These are the actin and myosin molecules. The overlapping and relative positions of these filaments determine the changes in force development. The "sliding filament hypothesis", as it is called, formed the basis for understanding the contraction process in striated muscle. Figure 5.1 describes this scheme. The tension initially increases with increasing length of the fiber, until the sarcomere reaches a length of about 2.2 μm; the tension reaches a maximum and then declines as the length is increased further. The overlapping cross bridges determine the force generation. This scheme is illustrated in Figure 5.2.

The relation between sarcomere length and force development has also been investigated in isolated segments of heart muscle. Mammalian papillary muscle preparations from cat and rat, for example, have been popular subjects for study. The longitudinal alignment of the papillary muscle fibers make them attractive for such analysis, especially for length measurements with optical or electron microscopes.

It is particularly interesting to note the fact that the elastic behavior of the sarcomere is such that its diastolic length is prevented from exceeding 2.3 μm. The sarcomere tension-length relationship has been translated to the intact global heart in terms of left ventricular developed pressure and end-diastolic volume. The former is referred to as the developed tension or force, while the latter is referred to as the

0-8493-0169-6/96/$0.00+$.50
© 1996 by CRC Press, Inc.

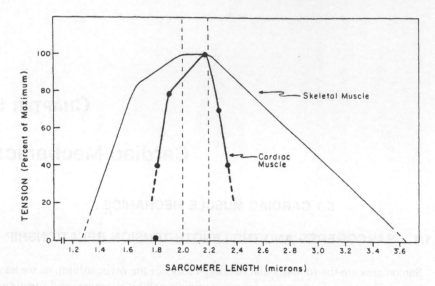

Figure 5.1   Sarcomere tension-length relationship. (From Sonnenblick, E. H., and C. L. Skelton, *Circ. Res.*, 35:517–526, 1974. With permission.)

initial muscle fiber length or "preload". More recently, it has been found that negative pressure is needed to empty the ventricle completely. This "ventricular suction" concept, though useful, has not been widely acknowledged. Figure 5.3 shows the relation between average midwall sarcomere length and filling pressure for dog and cat ventricles. Each sarcomere has corresponding similarities in structural and functional arrangements.

The relation between sarcomere length and left ventricular volume, however, is not a linear one. This is expected as the left ventricle is noncubical in shape. Thus, the percentage change in fiber length cannot be interpreted as a corresponding percentage change in left ventricular cavity volume. Even in the isovolumic ejection phase, when the ventricular volume remains constant, muscle fiber lengths measured at different parts of the ventricle may still be changing. Thus, isovolumic (constant volume) contraction does not exactly correspond to isometric (constant length) contraction. The relationship of regional tension development and global ventricular pressure increase is seen to be parallel however, only during this phase (Figure 5.4; Li, 1987). The anisotropic properties and differential epicardial and endocardial segmental contraction due to their respective fiber orientations complicate the direct translation of mechanics from the muscle level to the global ventricular level.

5.2 HILL'S VIEW OF FIBER MECHANICS

Hill was concerned about the mechanical efficiency, in terms of work and speed, of human muscles (1922, 1970). Although the concept of a mechanical spring as an energy storage element was introduced to model muscle behavior before him, Hill

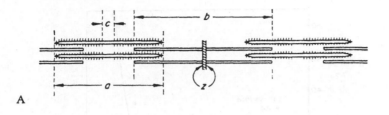

A

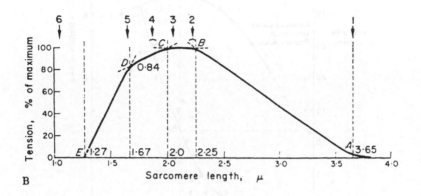

B

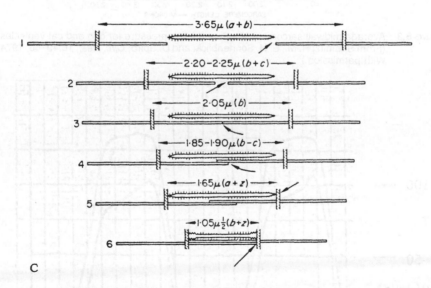

C

Figure 5.2 Illustration of the role of the cross bridges in sarcomere tension development. (From Gordon, A. M., A. F. Huxley, and F. J. Julian, *J. Physiol.*, 184:170–192, 1966. With permission.)

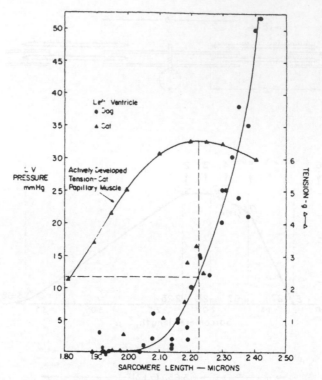

Figure 5.3   Average midwall sarcomere length and filling pressure for dog and cat ventricles. (From Spotnitz, H. M., E. H. Sonnenblick, and D. Spiro, *Circ. Res.*, 18:49–66, 1974. With permission.)

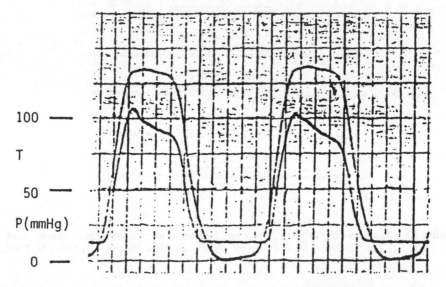

Figure 5.4   Simultaneously measured regional tension and global ventricular pressure in a canine left ventricle. Notice the parallel relationship in the isovolumic phases.

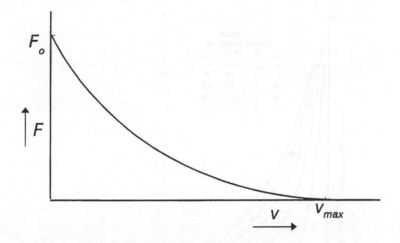

Figure 5.5 The inverse relationship of the force and velocity expressed by Hill's equation.

accounted for the energy dissipation through the introduction of a viscoelastic model. This model has an elastic spring, which can develop shortening, and a viscous dashpot element, which opposes shortening. This leads to the expression

$$K(1-l_o) - \beta \, dl/dt = F \tag{5.1}$$

where l is the muscle length and l_o is the initial muscle length, and the velocity of shortening is represented by the rate of change in fiber length

$$v = dl/dt \tag{5.2}$$

This velocity term thus gives rise to the viscous effect. At any muscle length l, a larger load F is lifted with a lower velocity than a smaller load. Thus, the force and velocity of the contractile element has an inverse relationship. This force-velocity relation is illustrated in Figure 5.5.

Hill's two-element model, consisting of a passive series elastic element, and a contractile element, has become popular and follows the general expression

$$(F+a)(v+b) = k \tag{5.3}$$

where k is a constant. Referring to Figure 5.6, F_0 is the maximum force developed by the isometrically contracting muscle. The velocity of shortening is a function of initial length of the muscle fiber, as Sonnenblick et al. (1964) found; this is also shown in Figure 5.6.

Combined with the earlier force-length relation, the force-velocity-length relation has been suggested to completely describe the physical behavior of the muscle.

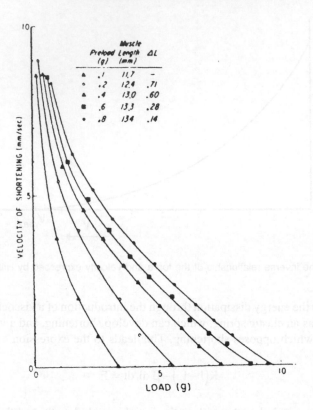

Figure 5.6   The force-velocity relation of the contractile element of the papillary muscle of a cat.
With increased intial fiber length, the velocity of shortening increases with a given
load. The maximum velocity of shortening (V_{max}) occurs at zero load. (From
Braunwald, E., J. Ross, Jr., and E. H. Sonnenblick, *Mechanisms of Contraction of
the Normal and Failing Heart*, Little, Brown, 1976. With permission.)

5.3 STARLING'S LAW APPLIED TO THE MAMMALIAN HEART

The different phases of global ventricular and atrial contraction defined from
simultaneous pressure and volume measurements are shown in Figure 5.7. The
systole comprises the isovolumic contraction phase or the pre-ejection phase, the
maximal ejection phase, and reduced ejection phases. The diastole consists of the
early relaxation, the rapid filling, and the reduced filling phases. It is clear that the
pressure-volume relationship defines much of the mechanical performance of the
heart.

5.3.1   STARLING'S LAW

Cardiac function is governed by several important physiological laws. Among
these is the well-known Starling's law of the heart. In his Linacre lecture, Starling
(1918) illustrated the importance of the input-output relationship of the heart. He
pointed this out when stating that the "energy of contraction is proportional to the

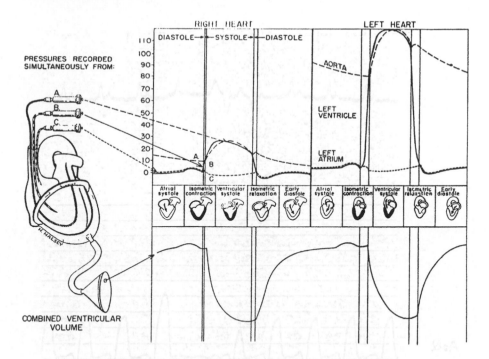

Figure 5.7 The various phases of cardiac contraction. (From Rushmer, R. F., *Structure and Function of the Cardiovascular System*, W. B. Saunders, Philadelphia, 1972. With permission.)

initial length of the cardiac fiber". This has come to be known as Starling's law of the heart. This law states that the beat-to-beat force-generating capacity of the intact ventricle is a function of its initial size, or the end-diastolic volume (EDV). This law is also widely referred to as the Frank-Starling mechanism, since Frank had earlier demonstrated this in the frog heart. At the fiber level, it dictates that the force of contraction or the extent of muscle shortening is dependent on its initial muscle length. This also means that the extent of muscle shortening is dependent on the end-diastolic length of the cardiac muscle fiber. At the global ventricular level, this law indicates that the ejected volume or the stroke volume (SV) is dependent on the end-diastolic volume. In terms of energy generation, the external work (EW), a product of mean arterial pressure and stroke volume, is also dependent on the preload.

The active tension-length relationship at the fiber level and the pressure-volume relationship at the global ventricle level are thus corresponding and important considerations. The initial fiber length and end-diastolic volume, or EDV, are generally defined as "preload" to ventricular ejection.

As discussed earlier, the length-tension relationship in cardiac muscle is such that as the muscle is stretched, the developed tension increases to a maximum, and then it declines as stretch becomes more extreme. The relation between ventricular performance and end-diastolic volume is brought about through the operation of the Frank-Starling mechanism, as demonstrated in Figure 5.8. A characteristic relation-

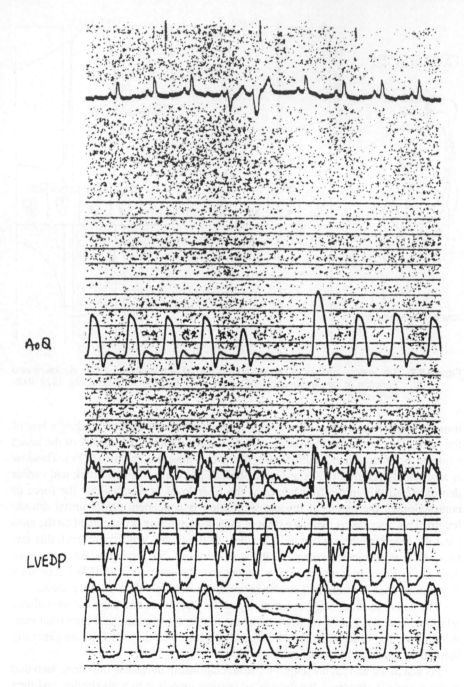

Figure 5.8   An illustration of the operation of the Frank-Starling mechanism applied to mammalian hearts.

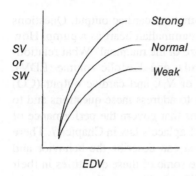

Figure 5.9 Left ventricular function curve with the stroke work plotted against filling pressure, for normal, strong, and weak hearts.

ship has been demonstrated between the velocity of shortening and the force development. It has been shown that increasing initial muscle length increases the maximally developed force without changing the maximal velocity of shortening (v_{max}). For this reason, v_{max} has been used as an index of contractility of the cardiac muscle.

5.3.2 LIMITATIONS OF STARLING'S LAW

The maximum tension developed by the sarcomere observed experimentally at a length of 2.2 μm corresponds to about 11.7 mmHg for the mammalian left ventricular chamber pressure. This is significant in differentiating cardiac function between normal hearts and those with disease conditions such as hypertrophy or the thickening of the ventricular wall. In a diseased heart, the end-diastolic volume and pressure can be considerably increased above 12 mmHg. In the case of coronary arterial disease, an end-diastolic pressure exceeding 30 mmHg is not uncommon. When ventricular function is depressed, stroke volume can be adequately preserved by progressive augmentation of the end-diastolic volume. With acute augmentation of diastolic filling pressure and volume, sarcomere length is increased until it approaches its limit of 2.2 μm. With sustained pressure and volume overload, however, despite the augmented volumes and filling pressures, sarcomeres do not display an enlarging H zone with a decrease in overlap between thick and thin filaments, and the 2.2 μm length is maintained. Thus, a further increase in stroke volume cannot be achieved. When the left ventricular function curves (Figure 5.9; Sarnof et al., 1965) are introduced, the descending limb of the curve cannot be obtained in experimental studies. It is not clear whether such impaired sarcomere function is reversible or replaceable.

It should be made clear that the Frank-Starling mechanism does not work if the mechanism involves a change in contractility as in the case of pulsus alternans (Li, 1982). As we shall see in Chapter 8, the Starling mechanism is an input-output relationship, displaying an open-loop control.

5.3.3 ALLOMETRIC RELATIONS

The mammalian heart varies greatly in size and rate, from 650 beats per minute in the 0.1-g heart of a small mouse to 15 beats per minute in the heart of a large whale.

Similar variation would also exist for stroke volume and cardiac output. Questions arise concerning the functional difference of the mammalian heart as a pump. How does the function differ with the widely varying size of the mammal? What relationships are there between heart rate and left ventricular end-diastolic volume (EDV), end-systolic volume (ESV), stroke volume (SV or V_s), and cardiac output (CO) among mammals of different size? It is necessary to address these questions and to examine the various laws and quantitative relations that govern the performance of the heart. We shall look at Starling's law here and Laplace's law in Chapter 7. There are many measurable and derivable quantities that can describe the behavior and function of the heart as a pump. We shall examine some of these quantities in their allometric forms to see whether there are similar working mechanisms among mammalian hearts.

The end-diastolic volume (EDV) in allometric relation is

$$EDV = 1.76W^{1.02} \tag{5.4}$$

and that of the stroke volume (SV) is

$$SV = 0.74W^{1.03} \tag{5.5}$$

It is clear that the end-diastolic volume increases with increasing body weight. This is mainly because larger mammals have larger hearts. Left ventricular pressures, however, are very much maintained constant, leaving the end-diastolic pressures and also the filling pressures of very similar magnitudes. The larger stroke volume of a larger mammal is therefore a result of a larger end-diastolic volume, not of a higher filling pressure. Starling's law thus operates both intra- and interspecies. It is interesting to note here that end-diastolic pressures in mammals are similar, despite differences in end-diastolic volumes.

With the consequence of a changing heart rate and a changing stroke volume in mammals, the cardiac output is modified. The heart rate, as shown previously, is given by (3.22)

$$f_h = 4.02W^{-0.25} \tag{5.6}$$

Cardiac output can then be easily obtained, neglecting right atrial pressure, as before,

$$CO = SV \times f_h \tag{5.7}$$

For a 70-kg man, this allometric equation gives a value of about 5 l/min, a very reasonable estimation. Thus, for a given heart rate, Starling's law allows adjustment of cardiac output by adjusting end-diastolic volume so as to provide a desirable stroke volume and energy generation (EW).

**Table 5.1 Ejection Fractions in Some
Adult Mammalian Species**

Man	0.67
Dog	0.65
Cat	0.64
Rabbit	0.61

5.4 SIMILAR EJECTION FRACTION AND CONTRACTILITY OF THE HEART

5.4.1 ALLOMETRIC RELATIONS

The amount of blood ejected out of the ventricle per beat, or stroke volume, is dependent not only on the operating Frank-Starling mechanism, or the preload, but also on afterload, contractility, and heart rate. Starling's law can be expressed, in part, in terms of the fractional volume that is ejected to the total resting volume. The ejection fraction (EF) so defined as the ratio of stroke volume to end-diastolic volume, is practically a constant in mammalian hearts,

$$EF = SV/EDV \tag{5.8}$$

With the use of the above allometric equations, this becomes

$$EF = 0.64 \tag{5.9}$$

The normal mammalian heart has an ejection fraction reported to be between 0.5 to 0.7. The lower calculated ejection fraction may be due to experimental conditions, especially under anesthetized open-chest situations. Table 5.1 provides some of the reported values of ejection fraction in different species of mammals.

Clinically, ejection fraction has been widely used as an index of left ventricular performance. The popularity of ejection fraction as an index probably stems from the fact that it depends on the ratio of only two volumes expressible in a dimensionless number that is also invariant of body weight. In addition, it embodies, in part, Starling's law relating stroke volume to end-diastolic volume.

5.4.2 EJECTION FRACTION AND VENTRICULAR CONTRACTILITY

For a given end-diastolic volume and an ejected stroke volume, the residual volume remaining in the ventricle is the end-systolic volume,

$$ESV = EDV - SV \tag{5.10}$$

Thus, by lowering the end-systolic volume, the stroke volume can be further increased. This capability to change the residual volume ESV, by which the heart is able to increase SV, undergoes a mechanism not prescribed by Starling's law. This change in SV will also change the ejection fraction. Ejection fraction, however, is a parameter that results from the combined behavior of ESV and EDV for each heart

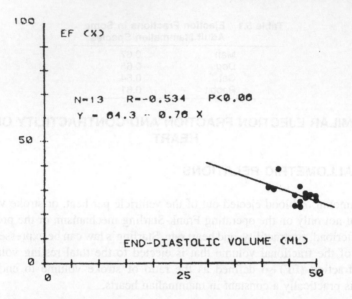

Figure 5.10 Ejection fraction plotted against end-diastolic volume. (From Kerkhof, P. L. M., *End-Systolic Volume and the Evaluation of Cardiac Pump Functions*, University of Utrecht, 1981. With permission.)

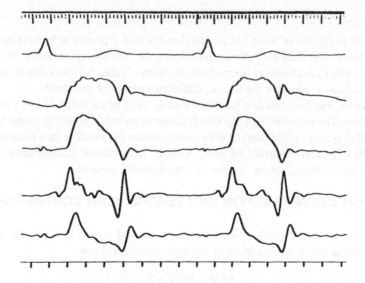

Figure 5.11 Ejection fraction plotted against end-systolic volume. (From Kerkhof, P. L. M., *End-Systolic Volume and the Evaluation of Cardiac Pump Functions*, University of Utrecht, 1981. With permission.)

beat. When EDV remains constant, the relation between EF and ESV is by definition a linear one, as seen from Equation (5.8). The dependence of EF on ESV and EDV for various conditions has been studied by Kerkhof (1979) and is schematically illustrated in Figures 5.10 and 5.11.

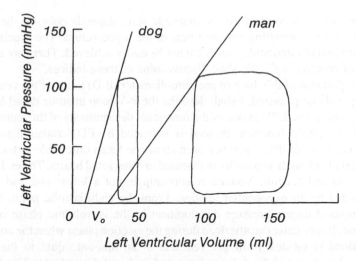

Figure 5.12 Left ventricular pressure and volume diagram of the mammalian heart. Pressure magnitudes are generally similar, while volume magnitudes are proportional to body weights.

Reducing ESV will always lead to an increase of ejection fraction, regardless of whether EDV or SV is constant. However, an increase of EDV will lead to an increase of ejection fraction when ESV is constant, but not when SV is constant. Since the physiological range of SV for each mammal is limited, as seen from the allometric relation, an increased end-diastolic volume will ultimately result in an augmentation of the end-systolic volume with a concurrent reduction of the ejection fraction.

5.5 THE PRESSURE-VOLUME CURVE AND CONTRACTILITY OF THE HEART

The Frank-Starling mechanism has classically been regarded as a fundamental property of the ventricle by which it regulates its cardiac output. As we have discussed, the larger the volume before the onset of contraction or preload, the higher its output under otherwise identical conditions. In this context, the preload has been identified as the end-diastolic volume. We have seen that this law holds true in terms of its behavior in isolated muscle preparations and in the intact ventricle. For this reason, stroke volume has long been used as an index of cardiac performance, since Frank's time, about a century ago.

The pressure-volume relation has been popular because it provides the interrelationships between SV, EDV, ESV, and EF on a single diagram. The pressure-volume diagram (Figure 5.12) is constructed from the instantaneous recordings of left ventricular pressure and volume. In addition, the separation of cardiac pump function and contractile state can be obtained. It is necessary, however, to distinguish between pump function and contractile state of the heart. Pump function is related to loading conditions, whereas contractility or inotropic state depends only on the intrinsic properties of the contracting muscle fibers or the myocardium. The precise quantitative definition of contractility itself cannot be easily achieved. There are numerous "indices of contractility". We shall discuss some of these indices.

The importance of P-V loop or pressure-diameter (P-D) loops of the ventricle has been frequently emphasized. I shall describe the P-V loop in some detail here. This diagram points to the ESV as one of the functional determinants of the representation of ventricular pump function. Preload is indicated by EDV rather than by end-diastolic pressure (EDP), which has been shown to be an unreliable index of initial myocardial fiber length, especially in diseased or innervated hearts. The end-diastolic pressure and end-diastolic volume relationship is not a linear one and has been approximated by an exponential relation. From the end-diastolic point, the curve extends upward during pressure development in the isovolumic phase of cardiac contraction. It encounters an afterload during the ejection phase when the aortic valve opens. Blood is ejected from the ventricle in an amount equal to the SV, the difference between end-diastolic volume and end-systolic volume. The trajectory reaches a peak and then declines slightly to the end-systolic point. The isovolumic relaxation phase begins and the pressure decays to a lower, but not a constant level and remains at about that level throughout the filling phase. The trajectory completes the loop at the end-diastole point. It is clear that each P-V loop represents a single heart beat. Thus, the beat-to-beat analysis can be accomplished with a family of such P-V loops.

In assessing the performance of the heart as a pump, it is necessary to understand the preload, afterload, and contractile states. In addition, heart rate is an important variable affecting these components. In other words, cardiac performance is dependent on heart rate, the filling and ejection characteristics, as well as on the intrinsic force-generating capability of cardiac muscle, resulting in what is called the "inotropic state". The instantaneous size and shape of the heart, or geometric properties, are also important determinants.

Notice that the time constant of ventricular relaxation can be calculated from a monoexponential fit to the isovolumic ventricular pressure fall from end-systolic pressure to the start of rapid filling,

$$p_v(t) = p_{ves}\, e^{-t/\tau_v} \tag{5.11}$$

As a general definition, afterload can be considered to be the force that resists ejection of the ventricular outflow. A more detailed description of afterload is discussed in the following chapter, since afterload is intimately related to the properties of the arterial system.

5.5.1   VENTRICULAR ELASTANCE

The pressure-volume relationship can be used to separate the effects of preload, afterload, and contractile state. In the excised-supported canine left ventricle and in normally ejecting hearts, the end-systolic points of a family of P-V curves have been shown to fall almost on a straight line. In other words, this line is tangent to the end-systolic point of a given pressure-volume curve. The ventricular elastance is then defined by

$$E(t) = p_v(t)/[V(t) - V_o] \tag{5.12}$$

where t denotes time, and V_o is a constant, known as the dead volume, which varies among mammalian hearts. The time course of this elastance is shown and discussed in Chapter 8. The maximal value of E(t), or E_{max}, represents systolic elastance, at end-systole,

$$E_{max} = ESP/(ESV - V_o) \tag{5.13}$$

This maximal elastance of the ventricle has been used as an index of contractility relatively independent of preload and afterload. With an increase in the contractile state, the slope of this line is shown to be steeper. Conversely, the slope declines with a decreased contractile state. A parallel shift of the end-systolic P-V curve will result in a change of V_o, thus indicating an altered contractility. This aspect does not conform to the original definition where V_o is assumed a constant. The rightward shift is now recognized as symbolic of a depressed heart. The significance of V_o has not yet been meaningfully explained. In general, V_o increases with increasing mammalian heart size, while E_{max} is larger in smaller mammals. This latter can be normalized with cardiac muscle volume, or left ventricular weight.

More recently, we have shown that the ventricular elastance variation through the cardiac cycle has its temporal relationship to the arterial elastance (Li, 1993). This aspect is important in the optimal operating performance of the cardiovascular system (Chapter 8).

5.5.2 EXTERNAL WORK AND ENERGY

Another important aspect of the pressure-volume curve is the area under the P-V loop, which represents the mechanical work performed by the ventricle to overcome its load. For this reason, it is often termed the external work of the heart, and the area under the loop is also known as the *work loop*. It is clear that this area is approximately the product of mean aortic pressure and the stroke volume,

$$EW = p \times SV \tag{5.14}$$

Thus the external work is proportional to body weight, in the same manner as the stroke volume. This is expected, since a larger mammalian heart will generate more external work. We shall elaborate this in more detail in Chapters 7 and 9.

CHAPTER 6

Arterial System Function

6.1 RHEOLOGICAL PROPERTIES OF MAMMALIAN ARTERIES

Structural composition and functional behavior of mammalian arteries vary according to anatomical sites. It is, however, useful to examine their mechanical properties in terms of the role of the arterial system in overall circulatory function.

6.1.1 ELASTIN, COLLAGEN, AND SMOOTH MUSCLE MECHANICAL PROPERTIES

Within the arterial wall, collagen is the stiffest wall component, with an elastic modulus of 10^8 to 10^9 dyn/cm^2 (Burton, 1954). This is some two orders of magnitude larger than those of elastin, 1 to 6×10^6 dyn/cm^2, and smooth muscle, 0.1 to 2.5 $\times$ 10^6 dyn/cm^2. We should note here that stiffness is proportional to elastic modulus; a larger elastic modulus means that the blood vessel is stiffer. Elastin is thus relatively extensible. Collagen, on the other hand, is relatively inextensible.

There are considerable data on the physiology of vascular smooth muscle (e.g., Somlyo and Somlyo, 1968). Smooth muscle can exert influence on large blood vessels such as the aorta. Its activity in smaller arteries is greater, because of the increased wall thickness (h) to radius (r) ratio (h/r). The mechanical properties of arteries are largely influenced by the behavior of smooth muscle. Elastic properties of the latter are dependent on the degree of activation. Smooth muscle activation alters the dynamic or frequency-dependent elastic modulus to a great extent. Mechanical properties of arterial vessel walls can also be altered by neural receptor mechanisms and by circulating catecholamines, such as norepinephrine.

The wall components operate to complement one another. For instance, at low blood pressure levels, elastin dominates the composite behavior, and at high pressures, collagen becomes more important. Elastic modulus increases with increasing blood pressure, and is nonlinearly dependent on the level of blood pressure (Li et al., 1990, 1993; Li and Zhu, 1994).

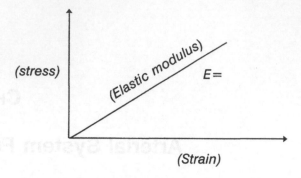

Figure 6.1 Stress-strain relation illustrating Hooke's law of elasticity. The slope gives the Young's modulus of elasticity.

6.1.2 ELASTIC MODULUS OF BLOOD VESSELS

Young's modulus of elasticity (E) is defined as the ratio of tensile stress (σ) to tensile strain (ε) for a material that obeys Hooke's law. It is valid only for application to blood vessels (Figure 6.1) when the deformation is small, i.e., under small strain conditions,

$$E = \sigma/\varepsilon \qquad (6.1)$$

where stress is force per unit area:

$$\sigma = F/A \qquad (6.2)$$

Notice that stress also has the unit of pressure. The higher the blood pressure, the greater the arterial wall stress. The tensile strain is extension per unit length:

$$\varepsilon = \Delta L/L \qquad (6.3)$$

For a blood vessel considered to be of a purely elastic material, Hooke's law applies and Laplace's law for wall tension can be written as

$$T = pr \qquad (6.4)$$

where p is the intramural-extramural pressure difference. This assumes that the blood vessel is thin walled or that the wall thickness to lumen radius ratio, h/r < 0.1. When the arterial wall thickness is not negligible as in a thick-walled vessel, a modified equation according to Lamé is used:

$$T = pr/h \qquad (6.5)$$

The Poisson ratio can be used to quantitatively express the compressibility of a biological material, such as that which composes arteries. It is defined as the ratio of

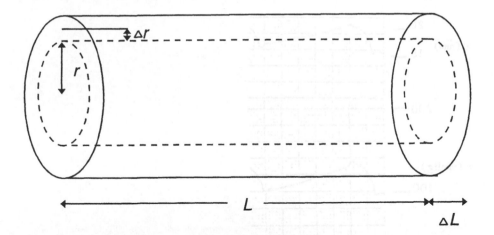

Figure 6.2 An arterial segment that exhibits radial and longitudinal strains. Their ratio defines tha Poisson ratio.

strain in the radial direction (distension) to the strain in the longitudinal direction (extension). This is illustrated in Figure 6.2.

$$\sigma_p = (\Delta r/r)/(\Delta L/L) \tag{6.6}$$

When $\sigma_p = 0.5$, the material is said to be incompressible. This means that the radial strain is twice the longitudinal one. This also says that when a cylindrical material is stretched, its volume remains unchanged. The wall of a blood vessel such as an artery is close to being incompressible; its Poisson ratio is about 0.48.

Along the arterial tree from the central aorta down to the peripheral arteries, or longitudinally, the number of elastic laminae decreases with increasing distance from the aorta, but the amount of smooth muscle increases and the wall thickness to radius ratio also increases. The vessel wall stiffness is thus increased. This latter phenomenon accounts for the large increase in the propagation velocity of the pressure pulse, commonly known as the pulse wave velocity (c).

It is interesting to note here that mammalian arteries tend to have similar elastic moduli at corresponding anatomical sites. Since blood pressures in mammals are similar and the pulsatile distension is similar, Young's modulus will likewise be similar. This conclusion can also be drawn from the relationship of the pulsatile distending pressure (ΔP) and the resulting diameter changes,

$$E = \Delta P/(\Delta D/D) \tag{6.7}$$

This formula is also known as the pressure (ΔP) — strain ($\Delta D/D$) elastic modulus. The simultaneous measurement of arterial blood pressure and diameter suffices to determine this modulus. An example of the recording of such a measurement is shown in Figure 6.3 where pressure is measured with a high-fidelity transducer and the diameter measured with piezoelectric ultrasonic dimension gauges.

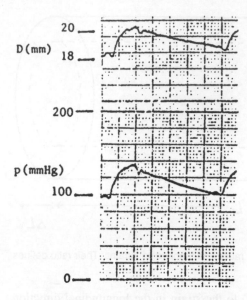

Figure 6.3 Simultaneously measured pressure and diameter in the ascending aorta of a dog.

The various components of the arterial wall are not arranged uniformly. Consequently, a constant elastic behavior across the arterial wall is not expected. For this reason, the arterial wall is not isotropic. A cross section along the longitudinal direction shows a helical organization of the collagen fiber network. It is this network that contributes mostly to the anisotropic properties of the arterial wall. This anisotropy also leads to the nonlinear blood vessel behavior. The anisotropy of the arterial wall has been shown from rheological studies, and the stress-strain relationship by Oka (1981), for instance.

6.1.3   VISCOELASTIC PROPERTIES OF BLOOD VESSELS

The static elastic properties of mammalian arteries have been extensively investigated. The stress-strain relationship, as discussed above, or the tension-length relationship has been shown to be nonlinear and thus does not obey Hooke's law, which only describes a purely elastic material. If the artery had been purely elastic, its modulus would increase linearly and the tension-length relation would be linear, as illustrated in Figure 6.1. Arterial elasticity increases with extension and the length-tension relation is curvilinear (Figure 6.4). The proximal aorta, which has little viscosity, has frequently been approximated to be purely elastic. A viscous material encounters losses or energy dissipation. A purely elastic material is lossless, by definition.

A purely elastic material differs from a viscoelastic material. The former depends only on strain (e), whereas the latter depends on the rate of stretch or strain rate ($d\varepsilon/dt$) also. The artery is viscoelastic. Most biological materials are characterized by their viscoelastic behavior. A viscoelastic material exhibits stress-relaxation, creep, and hysteresis phenomena (Figure 6.5). If a strip of artery is subjected to a step change in length, the stress or pressure will initially increase and then will decay to

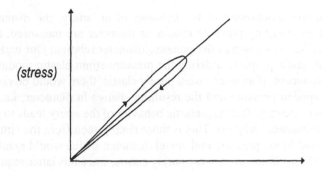

Figure 6.4 Tension-length or stress-strain relation of an artery. The straight line dictates the linear relation for a purely elastic material.

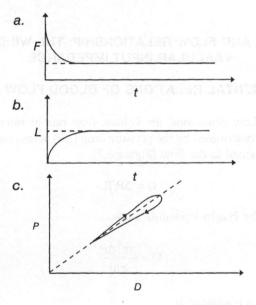

Figure 6.5 The characteristics associated with a viscoelastic material: (a) stress relaxation, (b) creep, and (c) hysteresis.

a lower value. This is known as stress relaxation ("relaxation lowers stress"). If it is subjected to a stepwise change in stress, its length will gradually increase to a constant value. This describes the creep phenomenon. These are the properties of an arterial wall subjected to transient or nonsteady state changes. Hysteresis develops when the vessel is subjected to sinusoidal or cyclic changes. When an artery is stretched with an applied force and then allowed to return to its original length, there is a loop formed which indicates the amount of energy loss due to hysteresis or energy dissipated through viscosity (resistance). A purely elastic material would return along the same exact path as when stretched.

For an *in vivo* assessment of the behavior of an artery, the distending blood pressure and the resulting pulsatile change in diameter are measured, as we mentioned earlier. This is known as the pressure-diameter relation. Our earlier measure of the arterial elastic property utilizing the pressure-strain elastic modulus is based on such measurement. If an artery were purely elastic, there would be no phase shift between the applied pressure and the resulting change in diameter, i.e., distension follows pressure exactly. The viscoelastic behavior of the artery leads to phase shifts in its pressure-diameter relation. This is more simply seen from the simultaneously measured arterial blood pressure and vessel diameter, which would exhibit identical pulsatile waveforms if the artery were purely elastic, since this latter requires proportional change in a linear fashion. It is clear from Figure 6.6 that this is not the case. The differences in pressure and diameter waveforms are more pronounced when blood pressure is elevated. Indeed, arteries have been shown to stiffen when pressurized.

6.2 PRESSURE AND FLOW RELATIONSHIP: THE WINDKESSEL AND VASCULAR INPUT IMPEDANCE

6.2.1   FUNDAMENTAL RELATIONS OF BLOOD FLOW

Under steady flow conditions, the volume flow rate of blood (Q) through an arterial segment is determined by the pressure drop (ΔP) across the segment and the resistance (R) presented to the flow (Figure 6.7),

$$Q = \Delta P / R \qquad (6.8)$$

It is governed by the Hagen-Poiseuille's law,

$$Q = \frac{\pi r^4 \Delta P}{8 \eta l} \qquad (6.9)$$

where the Poiseuille resistance is

$$R = \frac{8 \eta l}{\pi r^4} \qquad (6.10)$$

This resistance is defined for the steady flow or mean blood pressure and mean flow conditions. When fluid flow obeys such a relation (Equation 6.9), the fluid is said to be undergoing Poiseuille flow. This law was independently described by a German engineer, Hagen, and a French physician, Poiseuille, over a century ago. Under pulsatile flow conditions, this relation does not hold, and an impedance instead of resistance is defined, as we shall see in more detail later. It can be seen from this relation, however, that a change in lumen radius can most significantly alter the flow rate. This emphasizes the importance of having the right geometry or size.

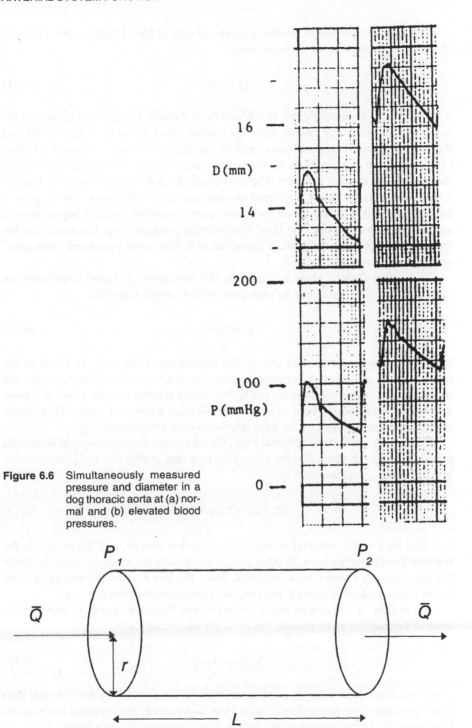

16 —

D (mm)

14 —

200 —

100 —

P (mmHg)

Figure 6.6 Simultaneously measured pressure and diameter in a dog thoracic aorta at (a) normal and (b) elevated blood pressures.

0 —

P_1

P_2

$\bar{Q}$ $\bar{Q}$

r

L

Figure 6.7 A cylindrical segment of an artery that is used to define the Poiseuille flow.

The volume flow rate is related to mean velocity of blood flow, or simply linear velocity (v) through cross-sectional area,

$$Q = vA \qquad (6.11)$$

It is clear from Poiseuille's law that blood flow through a large artery, such as the aorta, requires very little energy, since the viscous effect due to the wall is small, and that the large cross-sectional area presents negligible viscous resistance to flow. Consequently, it requires little mean pressure drop or gradient to propel the bulk of the steady flow through the aorta. The situation is much different in, say, the femoral artery, where lumen radius is reduced to some one fifth of the aorta. This requires a large differential in proximal and distal pressures, or in other words, a larger pressure gradient, in order to propel the flow. The sharpest pressure drop, however, is in the arterioles, which are predominant contributors to the "total peripheral resistance" seen by the mammalian heart.

With large vessels, such as the aorta, the movement of blood constitutes the inertial effect. This latter can be explained with a simple formula,

$$L = \rho l / A \qquad (6.12)$$

Opposite to the larger viscous effects, the movement of the mass of blood or the inertial effect in narrower peripheral vessels is much smaller. Consequently, the inertial effects in these vessels are negligible. Large arteries in general are also more distensible or more compliant (less stiff), as we have noted previously. Thus, larger arteries account for most of the total arterial system compliance (C).

For consideration of local blood flow, the more-than-two-century-old Bernouilli principle applies. It states that the sum of the potential energy and the kinetic energy is a constant, or mathematically

$$dP + \rho v^2 / 2 = \text{constant} \qquad (6.13)$$

The first term is the potential energy or the pressure energy, and the second is the familiar kinetic energy term. In other words, the greater the pressure drop, the faster the flow through a blood vessel segment. Thus, the loss in potential energy is made up by a gain in kinetic energy, obeying the conservation-of-energy law.

Blood flow is commonly characterized by the Reynolds number, which is the ratio of inertial forces to viscous forces as we discussed above:

$$Re = \rho v D / \eta \qquad (6.14)$$

where v is linear flow velocity and D is the lumen diameter. When Re is greater than 2000, turbulent flow rather than laminar flow is observed. This appears likely in the large aorta of a large mammal; whether it is true is discussed in Chapter 7.

6.2.2 PRESSURE AND FLOW WAVEFORMS IN ARTERIES

Pressure and flow in mammalian circulation are pulsatile in nature. Since modern catheterization techniques and electromagnetic flowmeters became available, blood pressure and flow waveforms have been measured in many mammalian blood vessels. Figure 6.8 gives an example of such waveforms at the outflow tract of the left ventricle, the ascending aorta of man, dog, rabbit, and guinea pig. Despite large body size differences and heart rates, the waveforms are similar.

Recordings of blood pressure waveforms, and of flow waveforms in different parts of the vascular tree in many mammalian species, have made clear some distinct features as the wave travels away from the heart: (1) the pulse pressure increases and the flow amplitude decreases progressively, although the mean pressure falls very slowly, except in the arterioles; (2) the rate of rise of the pressure wave increases and the wavefront becomes steeper, while the flow decreases in amplitude and rises more gently; and (3) the incisura produced by aortic valve closure becomes smoother and rounded off as the pressure wave propagates and the diastolic wave is accentuated. Figure 6.9 provides a graphic illustration of these points.

When pressure and flow waves encounter discontinuities, whether due to geometric taper, vascular branching, or elastic nonuniformity, reflections will occur and modify the propagating waveform. Amplification of the peripheral pressure pulse has been attributed to the presence of wave reflections, which we shall deal with in more detail in a later section. We have learned in Chapter 2 that the vascular wall becomes progressively stiffer accompanying reduced lumen diameters toward the periphery, which contributes to the pulse wave dispersion and, reflections.

The arterial wall and fluid viscosities attenuate the propagating pulse wave. Attenuation of the pulse waveform is also known as pulse damping. The attenuation is greater at higher frequencies. The incisura, for instance, has much higher frequency content than the rounded systolic wave, and hence, it is progressively more damped as it propagates. Indeed, when the pulse travels from the aorta to reach the femoral artery, the characteristic high-frequency features of the aortic pulse due to aortic valve closure have disappeared. Instead, a smooth waveform is seen. When the pulse reaches the arterioles, it is so damped that its waveform approaches sinusoidal. The arterioles have the largest drop in mean blood pressure. The arterioles, as mentioned previously, are responsible for the large peripheral resistance.

6.2.3 VASCULAR IMPEDANCES

The use of the Fourier series for the analysis of arterial function has been extensive. This is because of the periodic nature of pulsatile blood pressure and flow, and the assumed linearity of the arterial system. The Fourier series is indeed "a mathematical poem" with which one can see the beauty in the design features of the mammalian arterial trees. But as a poem, it requires background training in order to be appreciated. For this reason, its use has not been extensive in comparative biological studies. As we mentioned earlier, Fourier himself had discussed the importance of dimensional homeogeneity in the formulation of equations (Chapter

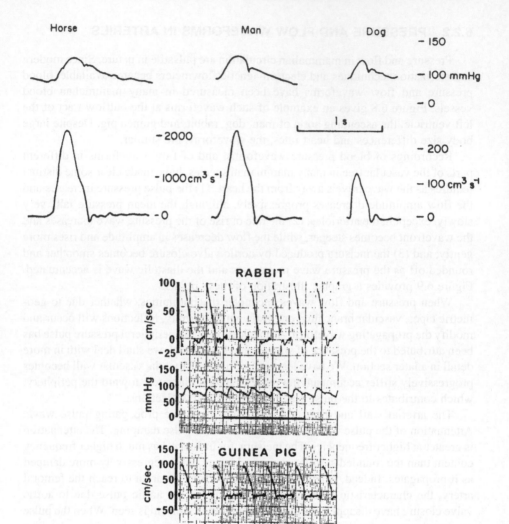

Figure 6.8 Pressure and flow waveforms at the root of aorta for four species of mammals. Notice the similarities. (Bottom figure from Avolio, A. P., M. F. O'Rourke, K. Mang and P. T. Bason, *Am. J. Physiol.*, 230:868–875, 1976. With permission.)

3). The use of Fourier analysis in the mammalian cardiovascular system was apparently introduced by Aperia (1940), but was first applied to analyze aortic pressure pulse by Porje (1946). With the advent of modern digital computers, it has now become a routine means of obtaining frequency domain information of the cardiovas-

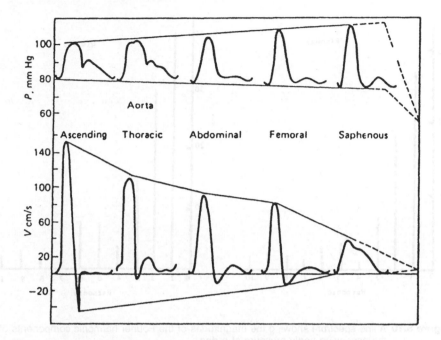

Figure 6.9 Simultaneously measured pressure and flow waveforms in different parts of the canine arterial tree. (From McDonald, D. A., *Blood Flow in Arteries*, Arnolds, London, 1960. With permission.)

cular system from time domain pressure and flow waveforms. An example that follows is the pulse transmission characteristics of the mammalian arterial trees. The basic principles of Fourier series and analysis have been discussed in Chapter 4.

Blood pressure and flow waveforms in the arterial system contain at the most ten (ten times the heart rate in beats/second) significant harmonics. In other words, these waveforms can be accurately reconstructed from the most, ten harmonic components or sinusoids. Figure 6.10 gives the magnitude spectrum of the ascending aortic pressure of a dog. In many cases, the magnitudes of the harmonics beyond the sixth is small and may subject to uncertainty, because of the noise level and the resolution of the recording instruments. A sampling rate of 100 Hz or sampling at every 10 msec is adequate for arterial pressure and flow waveforms (Attinger et al., 1966) Differential pressures and derivatives, however, contain much higher frequency components. These can be easily seen from the waveforms shown in Figure 6.11.

The input impedance of the arterial system is the harmonic ratio of measured ascending aortic pressure (p) to flow (Q),

$$Z = (P/Q)e^{j\theta} \qquad (6.15)$$

$$Z_n = |Z_n| \underline{/\theta_n} \qquad (6.16)$$

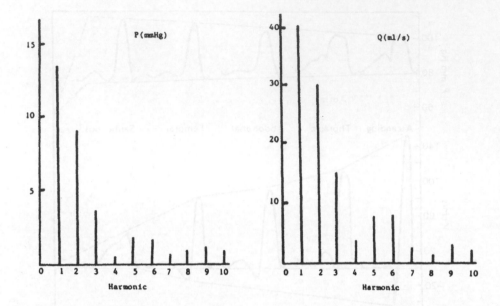

Figure 6.10 A line spectrum showing the magnitudes of the Fourier harmonic components of the ascending aortic pressure of a dog.

where P and Q = pressure and flow amplitudes, respectively, θ = phase difference between pressure and flow, and for the nth harmonic, $|Z_n|$ = modulus of impedance and θ_n = phase of impedance.

Measurement of the impedance of the vascular bed apparently was first attempted less than four decades ago. Impedances in the systemic and pulmonary arterial tree have been obtained by measuring flow (or computed flow from pressure gradient) and pressure at the ascending aorta and in some arteries of man and other mammals.

The input impedance of the systemic arterial tree represents the "load" presented to the heart during left ventricular ejection, or the impedance to blood flow that the left ventricle "sees". Thus, the determination of such impedance has important physiological implications. With higher impedances presented to the heart, ejection flow will decrease, unless the heart works harder to overcome this load.

Much of the impedance information is examined in man and dog more thoroughly than in other species. In these two mammalian species, the frequency dependence of input impedance shows a large decrease in magnitude at fundamental or very low frequencies (<2 Hz), then oscillates, exhibiting maxima and minima, and eventually reaches a somewhat constant level that is low compared to its zero frequency value (Figure 6.12). Zero frequency here refers to the nonoscillatory or mean pressure over mean flow component, or eqivalently the total peripheral resistance,

$$R_s = \overline{P}/\overline{Q} \qquad (6.17)$$

This assumes that the right atrial pressure is close to zero.

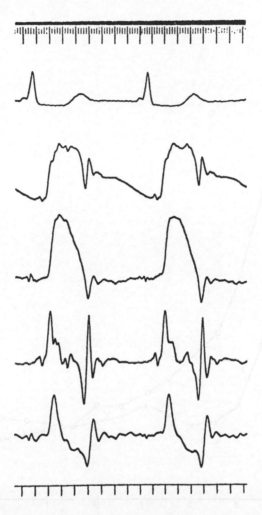

Figure 6.11 Simultaneously measured pressure and flow waveforms and their respective first derivatives. (From Li, J. K-J., *Arterial System Dynamics*, New York University Press, 1987. With permission.)

At higher frequencies (>5 Hz), the input impedance (Z) approaches the characteristic impedance (Z_o) of the proximal aorta. As the term implies, this is the impedance that is characteristic of the proximal aorta, and it is therefore dependent only on its geometric and viscoelastic properties. Alteration in properties of other arteries or vascular beds would not have any primary effects on this aortic characteristic impedance. Z_o can also be approximated from the water-hammer formula in the time domain,

$$Z_o = \rho c / A \qquad (6.18)$$

where A. as before. is the lumen cross-sectional area.

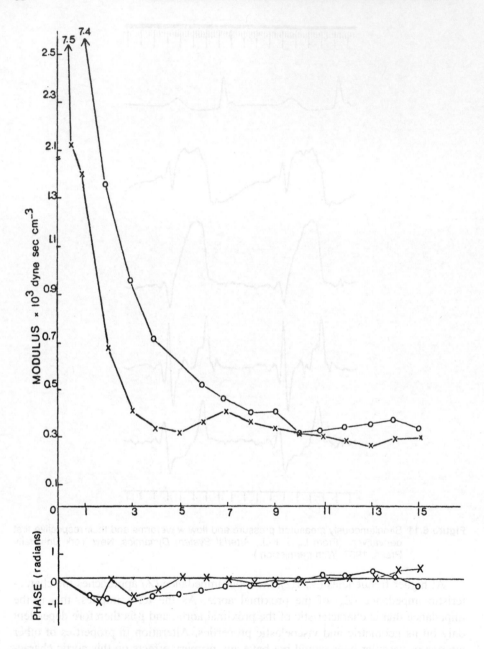

Figure 6.12 An example of the input impedance spectrum showing harmonic magnitudes and phases.

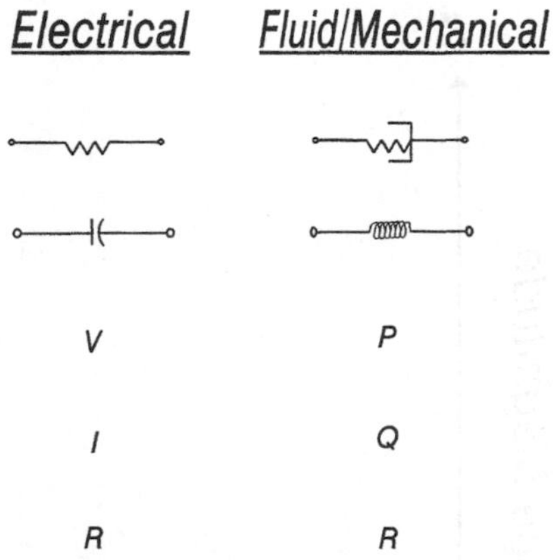

Electrical   Fluid/Mechanical

V	P
I	Q
R	R

Figure 6.13 Electric analogs of fluid mechanical parameters.

The phase of the impedance is initially negative, becoming progressively more positive with increasing frequency, crossing zero at about 3 to 5 Hz, and oscillating close to zero thereafter. For an electric analog of blood flow, fluid resistance is represented by electric resistance (R), blood flow is represented by electric current (I), blood pressure is electric voltage (V). Elasticity corresponds to capacitance (C), which stores energy. Figure 6.13 illustrates the analogy.

The simplest analog of the systemic arterial system is the windkessel model, consisting of a capacitor representing the compliant aorta and a resistor representing resistance of the stiff peripheral vessels. The model was originated by Frank (1899) in modeling the arterial system for the purpose of calculation of stroke volume from the aortic pressure pulse waveform. This model lacks pulse transmission characteristics, i.e., the pressure pulse appears everywhere in the arterial tree at the same instant, implying that the pulse is propagating at infinite wave velocity. Despite this limitation, the windkessel model is a good approximation to the behavior of the arterial system at low frequencies, in terms of input impedance (Figure 6.14). This model does not have an element representing the characteristic impedance of the proximal aorta. The three-element windkessel model is a refinement with an inclusion of Z_o (also shown on Figure 6.14).

With the availability of the windkessel model, Stahl (1963) and Kenner (1972) proposed the use of the ratio of windkessel time constant (τ) to the cardiac period (T) as a similarity criterion. The time constant is obtained from the monoexponential decay of the diastolic aortic pressure beginning at the end-systolic pressure and ending at diastolic pressure, or

$$P_d = P_{es}\, e^{-t/\tau} \tag{6.19}$$

and

$$K_t = \tau/t_d \tag{6.20}$$

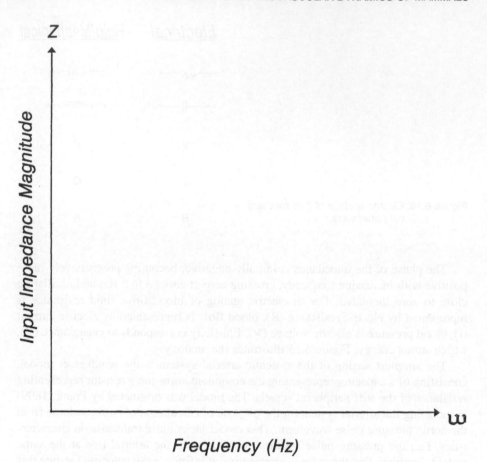

Figure 6.14 The regime of validity of the windkessel model as an approximation to the input impedance.

We have obtained values of K_t for several mammalian species and found K_t to be relatively constant. We shall also see in Chapter 7 that the ratio of characteristic impedance to the total peripheral resistance, or Z_o/R_s, represents the ratio of pulsatile to steady energy, as a dimensionless similarity criterion, i.e., a constant.

6.3 PULSE PROPAGATION WAVELENGTH AND ARTERIAL SYSTEM LENGTH

6.3.1 THE PROPAGATION CONSTANT, PULSE ATTENUATION, AND PULSE WAVE VELOCITY

For a pressure wave propagating along a uniform vessel without terminal reflections, the pressures measured simultaneously at any two points along the vessel are related by

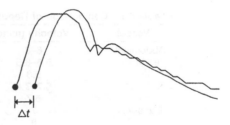

Figure 6.15 Method demonstrating how foot-to-foot velocity is obtained between two pressure-measuring sites in the arterial system.

$$C_f = \frac{\Delta z}{\Delta t}$$

$$p_2 = p_1 \, e^{-\gamma z} \tag{6.21}$$

where p_1 and p_2 are proximal and distal pressures, respectively, γ is the propagation constant, and z is the longitudinal coordinate in the direction of blood flow. The propagation constant describes the speed at which the pulse propagates and the manner in which the pulse is attenuated. For this reason, the propagation constant comprises the two (Li et al., 1981),

$$\gamma = \alpha + j\beta \tag{6.22}$$

where α is the attenuation coefficient describing the extent of the damping of the pulse, β is the phase constant. The pulse wave velocity or the phase velocity c is given by

$$c = \omega/\beta \tag{6.23}$$

In the presence of reflected waves however, the simultaneous measurement of p_1 and p_2 will not give the "true" wave velocity. Instead, the apparent phase velocity is obtained. The apparent phase velocity has a frequency dependence similar to that of the input impedance. Both are influenced by wave reflections. It approaches the "true phase velocity" at high frequencies. For a detailed discussion, the reader is referred to Li (1987).

A practical and simple approach to obtaining pulse wave velocity is to measure the arrival time of the "foot" of the two simultaneously measured pressure pulses. This is commonly known as the foot-to-foot velocity, which is easily calculated from the time delay (dt) and the distance between the pressure measurement sites (Δz),

$$c_{ff} = \Delta z/\Delta t \tag{6.24}$$

This method of obtaining pulse wave velocity is shown in Figure 6.15.

The foot-to-foot velocity thus gives a good approximation of the composite pressure wave at high frequencies. It is related to arterial wall elastic properties from the Moens-Korteweg formula,

$$c_o = \sqrt{Eh/2r\rho} \tag{6.25}$$

as described before.

Table 6.1   Comparison of Reported Wave Velocities

Vessel	Velocity (m/sec)	Data source	Method
Abdominal	6–8	Dog	f-f[a]
aorta	5.5–8.5	Dog	f-f
	9.3	Dog	dE[b], mv[c]
	6.7–7.4	Dog	f-f
	6.7	Dog	mv
Iliac artery	7	Young men	dE
	8	Old men	dE
	7–8	Dog	f-f
	7.7	Dog	mv
Femoral artery	8–12	Dog	f-f
	8	Man	f-f
	8.5–13	Dog	f-f
	9.3	Dog	dE, mv
	18	Young men	f-f
	13	Old men	f-f
	8 ± 1.1	Dog	mv
	8.3–10.3	Dog	f-f
	8.2	Dog	mv
	10.4 ± 0.4	Dog	dE
	8.5	Dog	mv
	8.8	Dog	mv
Carotid artery	5–12	Man	f-f
	9.4	Dog	dE, mv
	6.1–7.4	(Proximal) dog	f-f
	7.7–8	(Distal) dog	f-f
	8	Dog	mv

Source: Li et al. (1981). *Circ. Res.* 49:442–452.

[a] f-f, measured as foot-foot velocity.

[b] dE, calculated utilizing dynamic elastic modulus.

[c] mv, mean value from frequency spectrum of phase velocity.

The pulse wave velocity has a weak dependence on mammalian body weights. Its value at corresponding anatomic sites in the arterial tree thus varies little among species, as we shall see later. Measurements of pulse wave velocities along the pulse transmission path have shown increased wave velocity toward the periphery. Table 6.1 summarizes some of these measurements for different parts of the aorta as well as the carotid, iliac, and femoral arteries. Pulse wave travels more slowly in larger vessels, but suffers larger viscous damping in smaller vessels. These are related to the structure of the arterial tree.

6.3.2   ESTABLISHING THE RATIO OF PULSE WAVELENGTH TO ARTERIAL SYSTEM LENGTH

It is apparent from the above that different harmonic components of the pulse wave propagate with different speed and suffer from differential damping. Now we can relate the propagation characteristics to the structural design of the arterial system. We first notice that each harmonic component traveling with velocity c, and frequency f is associated with a wavelength λ. Thus, for a pressure pulse with 10 harmonics, there will be 10 different wavelengths. The distribution of wavelengths

Table 6.2 Hemodynamic and Anatomic Data for Different Mammals

	Body weight W (kg)	Heart rate f_h (beats/min)	Phase velocity c (m/sec)	System Length L_a (cm)	Wavelength λ(cm) Maximum	Wavelength λ(cm) Minimum	λ/L_a (dimensionless)	
Horse	400	36	400	110	667	55	6.0	0.5
Man	70	70	500	65	429	36	6.6	0.5
Dog	20	90	400	45	267	23	5.9	0.5
Wombat	16	120	500	40	250	21	6.2	0.5
Cat	3.6	180	450	27	150	13	5.6	0.5
Rabbit	3.0	210	450	25	129	11	5.2	0.4
Guinea pig	0.7	240	420	5	105	9	7.0	0.6
Mean			446				6.1	0.5

plays a key role in pulse wave transmission. Harmonic components with longer wavelengths can reach further along the transmission path than shorter ones, for instance.

On the basis of the expression

$$c = f\lambda \tag{6.26}$$

where c = phase velocity, f = frequency, and λ = wavelength, it is possible to estimate the range of values for the wavelengths. For example, in man taking for c a value of 500 cm/sec in the aorta and for the lowest frequency of interest $f_{min} = 1.2\,\text{Hz}$ (72 beats per min), the calculated maximum wavelength $\lambda_{max} = 429$ cm. For the 10th harmonic, or $f_{max} = 1.2 \times 10 = 12$ Hz, the wavelength is now 36 cm. These values can be compared to the arterial system length, or the length of the aorta, of 65 cm. Hence, the fundamental component of the pressure pulse can easily reach the terminal aorta ($\lambda = 429$ cm >> 65 cm to the aorto-iliac junction), whereas the 10th harmonic will not ($\lambda = 36$ cm < 65 cm). The distribution of the wavelengths of interest ranges from much longer to much shorter than the arterial system. The same situation applies to the canine and other mammalian arterial systems in which they occur, when allowance is made for the resting heart rate being faster than that in man (Li, 1978; Noordergraaf et al., 1979).

For the purpose of expressing this distribution of wavelengths as a nondimensional number, it is compared to a convenient measure of the arterial tree, namely, the length of the aorta from the aortic valve to the aorto-iliac branching junction (L_a in Table 6.2).

For the distribution of wavelengths relative to the size of the arterial system to qualify as an invariant property, it must apply to all mammals. Table 6.2 summarizes arterial properties for mammals ranging in size from the horse to the guinea pig with a greater than 570-fold differences in body weights. Accordingly, the distribution of the ratio of wavelength normalized by system length, adjusted for resting heart rate, was found to be a constant. Thus, there appears to exist a similarity principle that underlies the similarity in pressure and flow pulses observed in different mammalian species. The ratio between λ_{max} and L_s is essentially constant.

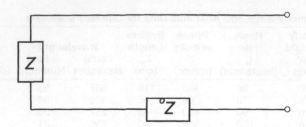

Figure 6.16 The uniform tube model of the arterial system with characteristic impedance Z_o, terminated in a load impedance Z.

6.4 PULSE WAVE REFLECTIONS

Although reflection of the pressure pulse wave has been observed a century ago, explanations of the mechanism of its occurrence began only in the last five decades or so. Borrowing from the principles of the propagation of sound waves, Hamilton and Dow (1939) claimed to have observed nodes and antinodes in the pressure pulses and attributed them to "standing waves". When the reflected wave peaks at exactly the same instant as forward traveling wave, they form the antinode; when the reflected wave is exactly opposite in phase, a node is formed. This theory holds if the arterial pulses were not attenuated. As we have shown, they are attenuated due to the viscosity of the arterial wall and of the fluid. Thus, standing waves do not necessarily exist.

We shall now discuss how to quantify the characteristics of pulse wave reflections in individual vascular beds, whether due to changing vasoactive states or mechanical interventions. For a single uniform blood vessel with characteristic impedance Z_o and terminated with vascular load impedance Z (Figure 6.16), the reflection coefficient Γ is given by

$$\Gamma = \frac{Z - Z_o}{Z + Z_o} \tag{6.27}$$

The reflection coefficient so obtained is therefore a complex quantity with magnitude and phase. The reflection coefficient thus varies with frequency. In other words, the amount of propagating pulse that is reflected for each harmonic component differs. We should note here that under steady pressure and flow conditions, there is no propagation of the pulse wave, hence, no reflections. For the whole arterial system, Z_o is identified as the characteristic impedance of the ascending aorta. Z is the input impedance of the systemic arterial tree. Thus, Γ is the global reflection coefficient. It is termed global because it does not differentiate its individual contributing sources, nor the effects of multiple reflections.

A local reflection coefficient can be defined, however, for a vascular branching junction. This is discussed in Chapter 9. Briefly, by examination of the characteristic impedances of the mother vessel and the branching daughter vessels, pulse wave reflection arising due to vascular branching is minimal (Li, 1984).

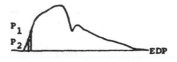

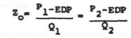

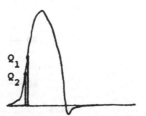

Figure 6.17 Time-domain estimation of the characteristic impedance of the aorta from instantaneous pressure and flow waveforms.

We shall now establish a method that can be used to resolve the measured pressure and flow waveforms into their forward and reflected components. Considering the pressure (p) and flow (Q) waveforms in the aorta as the sum of the forward (f) or antegrade and the reflected (r) or retrograde components, then simply,

$$p = p_f + p_r \tag{6.28}$$

$$Q = Q_f + Q_r \tag{6.29}$$

In order to resolve these components, the characteristic impedance of the aorta Z_o needs to be known. As we have discussed previously, Z_o is independent of reflections. In this manner, it can be defined in terms of forward and reflected waves,

$$Z_o = p_f/Q_f = -p_r/Q_r \tag{6.30}$$

Z_o can be obtained either from the water-hammer formula we introduced earlier or from the instantaneous ratio of the measured aortic pressure and flow waveforms (Figure 6.17; Li, 1986). This latter is true only in the very early part of the ventricular ejection, when reflected waves do not have sufficient time to return to the heart. Consequently, Z_o is given as the ratio of the aortic pressure to flow, above the end-diastolic level,

$$Z_o = \Delta p/\Delta Q$$
$$= (p - p_d)/Q \tag{6.31}$$

Again, this is a good approximation, particularly in the early ejection phase (the first 60 msec or so, shorter in smaller mammals due to shorter cardiac periods), when the amount of peripheral reflections reaching the proximal aorta is minimal.

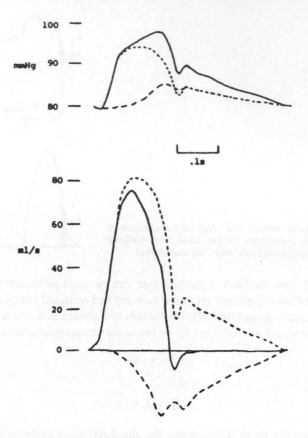

Figure 6.18 Measured aortic pressure waveform resolved into its forward and reflected components.

These expressions (Equations 6.28 to 6.31) permit the subsequent resolution of the forward and reflected waveforms,

$$p_f = (p + QZ_o)/2 \qquad (6.32)$$

$$p_r = (p - QZ_o)/2 \qquad (6.33)$$

An example of the resolved forward and reflected components of aortic pressure is shown in Figure 6.18.

Clearly, the reflection coefficient can be alternatively obtained as

$$\Gamma = P_r/P_f \qquad (6.34)$$

This is equivalent to the reflection coefficient that is obtained from characteristic and input impedances (Equation 6.27). It also varies with frequency or that the amount of reflections for different harmonic components of pressure and flow are different.

PULSE REFLECTION

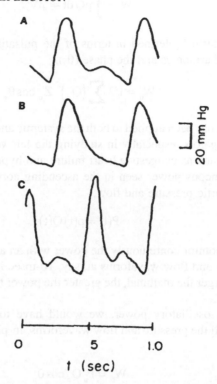

Figure 6.19 Femoral arterial pressure waveforms measured under (A) normal, (B) vasodilated, and (C) vasoconstricted states. (From Li, J. K-J., J. Melbin, and A. Noordergraaf, *Am. J. Physiol.*, 247:H95–H99, 1984. With permission.)

Wave reflections can modify the pulse waveforms appreciably. An example is shown in Figure 6.19 for the femoral arterial pulse during vasodilation and vasoconstriction. This is paticularly true for small arteries, where smooth muscle activation and vasoactive effects are predominant.

6.5 PULSATILE POWER GENERATION AND ENERGY DISSIPATION

The heart as an energetic source generates the pressure and flow pulses. Hydraulic power realized in the vascular system is made up of a steady, nonpulsatile term and a pulsatile or an oscillatory term. We can express this as

$$W_t = W_s + W_o \qquad (6.35)$$

The steady power can be calculated from the product of pressure and stroke volume or the integral of the product of pressure and flow over a cardiac period, as we discussed previously,

$$W_s = \int pQ\, dt \simeq pV_s \tag{6.36}$$

The oscillatory term is defined in terms of the pulsatile flow (Q_n) and the input impedance (Z_n) harmonics and the phase (θ_n),

$$W_o = 1/2 \sum (Q_n)^2 Z_n \cos\theta_n \tag{6.37}$$

This concept has been applied to both the systemic and the pulmonary circulation by many investigators, especially in studying the left ventricular energetics in the diseased states such as congestive heart failure and hypertension.

The instantaneous power seen in the ascending aorta is simply the product of instantaneous aortic pressure and flow,

$$P(t) = p(t)Q(t) \tag{6.38}$$

Thus, one can monitor continuously the power with an analog amplifier, or one can digitize pressure and flow waveforms at, say, 10-msec intervals and calculate their products. The larger the mammal, the greater the power that can be generated by the heart.

To compute oscillatory power, we would have to perform Fourier analysis (Chapter 4) on all the pressure and flow waveforms. In-phase pulsatile power is then defined by

$$W_n = P_n Q_n \cos\theta_n \tag{6.39}$$

where n represents the nth harmonic component and θ_n, its phase angle.

6.5.1   THE RELATIONSHIP OF POWER AND INPUT IMPEDANCE

We can use the electric analogy shown earlier to illustrate the effects on energy usage and storage properties. Power dissipates through resistors. Thus, the resistive element in the windkessel absorbs power. A capacitor can store energy. Thus, an elastic vessel wall actually stores the energy due to pulsations during systole and then returns it to the system by elastic recoil in diastole. This is a simple illustration of the "energy return system" or ERS. An inductor can generate energy only if there is changing current or, in our case, changing flow. The greater the instantaneous blood flow out of the left ventricle, the greater the inertia.

DC component or the total peripheral resistance R_s is the ratio of mean pressure to mean flow. It is significantly higher than the oscillating components. Therefore, the heart appears to be decoupled from its peripheral load, so that external cardiac work becomes remarkably independent of heart rate at high frequencies.

Ventricular afterload is defined as all external factors that oppose ventricular ejection. Arterial input impedance has been suggested as the afterload. This appears to be insufficient. The interaction of the afterload and inotropic state of the heart also remains unresolved. The energy design of the human circulatory system has been of much interest for many decades (Evans and Matsuoka, 1915; Rodbard et al., 1959; and Porje, 1967, for instance). The slower the heart rate, the greater the external work needed to eject a given pulsatile flow (Li, 1983). Given a constant impedance spectrum, however, the smaller the pressure and flow generated by the ventricle, the lower the external work; see Equation 6.36. In any case, it is important to note that both the ability of the left ventricle to do work (myocardial performance) and the loading properties of the arterial system are important in determining the power generated by the ventricle.

CHAPTER 7

Similarity Analysis of Cardiovascular Function

7.1 LAMINAR AND TURBULENT FLOW IN MAMMALIAN ARTERIES

We have discussed the pattern and characteristics of blood flow in arteries. In this section, we shall look at the fluid mechanical factors that govern such flow in the aortas of mammals from a comparative point of view.

Ventricular outflow attains its maximal amplitude during the early part of the ejection phase. The size of the outflow tract at the root of the aorta at any given velocity determines the amount of flow. The viscous drag, however, specifies the degree of retardation of this flow. The fluid dynamic relation of a Newtonian incompressible fluid and the viscous force against which the fluid flows were first derived over a century ago by Reynolds (1883). The Reynolds number describes steady fluid flow through a rigid pipe. It was soon employed to characterize the transition from laminar to turbulent flow in elastic tubes, but much later in mammalian arteries (e.g., McDonald, 1960; Oka, 1981). The occurrence of turbulence is of particular interest to hemodynamic studies because it can lead to detrimental effects on cardiovascular function. On a related concern, turbulence is also energetically wasteful.

Comparative studies of the mammalian cardiovascular system with respect to the fluid dynamic aspects of the circulation have not been extensive. We have seen that there are several means to derive circulatory similarity criteria of blood flow. They have been based mostly on the well-known hydrodynamic principles. The use of fluid mechanic criteria such as Reynolds number, though restricted in defining global hemodynamic similarities, offers insights into the local flow regime of the mammalian circulation. This section specifically examines the familiar Reynolds number in the aortas of selected species of mammals from a comparative point of view.

The combination of dimensional analysis, allometry, and hemodynamic principles has been shown to present a powerful method for establishing similarity principles (Li, 1983). For such analysis, allometric equations of parameters or variables under consideration need to be known first. Table 7.1 lists some relevant

0-8493-0169-6/96/$0.00+$.50
© 1996 by CRC Press, Inc.

Table 7.1   Allometric Relations of Some Hemodynamic Parameters
             ($Y = aW^b$, W in kg)

Parameter	Y	a	b	Ref.
Heart rate (sec^{-1})	f_h	3.60	−0.27	Adolph (1949)
Stroke volume (ml)	V_s	0.66	1.05	Holt et al. (1968)
Pulse velocity (cm/sec)	c	446.0	0.0	Li (1987)
Arterial pressure (dyn/cm^2)	p	1.17×10^5	0.033	Gunther and Guerra (1955)
Radius of aorta (cm)	r	0.205	0.36	Holt et al. (1981)
Length of aorta (cm)	L	17.5	0.31	Li (1987)

hemodynamic parameters for the present investigation in their allometric form. Weight (W) is expressed in kilograms.

A dimensional matrix is first formed by incorporating parameters that are pertinent to the current analysis. These are the fluid density (ρ) and viscosity (η), diameter (D) of the blood vessel, velocities of the flowing blood (v) and of the pulse wave (c). In terms of the dimensioning mass (M), length (L), and time (T) system, we can write down the following dimensional matrix,

	ρ (g/cm^3)	c (cm/sec)	D (cm)	η (P)	v (cm/sec)	
M	1	0	0	1	0	
L	−3	1	1	−1	1	(7.1)
T	0	−1	0	−1	−1	
	k_1	k_2	k_3	k_4	k_5	

where k_ns are Rayleigh indices referring to the exponents of the parameters. According to Buckingham's (1915) Pi-theorem (Li, 1983, 1986) two dimensionless Pi-numbers (5 − 3 = 2) can be deduced.

Mathematically, we have

$$\pi_i = \rho^{k1} c^{k2} D^{k3} \eta^{k4} v^{k5} \qquad (7.2)$$

or in terms of M, L, and T,

$$\pi_i = \left(M^{k1} L^{-3k1} T^0\right)\left(M^0 L^{k2} T^{-k2}\right)$$
$$\left(M^0 L^{k3} T^0\right)\left(M^{k4} L^{-k4} T^{-k4}\right)\left(M^0 L^{k5} T^{-k5}\right) \qquad (7.3)$$

This can be arranged to give

$$\pi_i = M^{(k1+k4)} L^{(-3k1+k2+k3-k4+k5)} T^{(-k2-k4-k5)} \qquad (7.4)$$

Since pi-numbers are dimensionless, this means the exponent needs to be zero. To achieve $M^0L^0T^0$, we need to have

$$k1 + k4 = 0$$

$$-3k1 + k2 + k3 - k4 + k5 = 0 \qquad (7.5)$$

$$-k2 - k4 - k5 = 0$$

There are three equations with five unknowns. Therefore, we shall solve k3, k4, and k5 in terms of k1 and k2.

$$k1 = 1\,k1 + 0\,k2$$

$$k2 = 0\,k1 + 1\,k2$$

$$k3 = 1\,k1 + 0\,k2 \qquad (7.6)$$

$$k4 = -1\,k1 + 0\,k2$$

$$k5 = 1\,k1 - 1\,k2$$

Hence the two pi numbers or similarity criteria are obtained from the solution matrix (Li, 1983):

	ρ (g/cm³)	c (cm/sec)	D (cm)	η (P)	v (cm/sec)	
M	1	0	0	1	0	
L	-3	1	1	-1	1	(7.1)
T	0	-1	0	-1	-1	
	k_1	k_2	k_3	k_4	k_5	

or

$$\pi_1 = \rho v D / \eta = Re \qquad \pi_2 = c/v = 1/Ma \qquad (7.8)$$

The Reynolds number is clearly identified as Re, and Ma is known as the Mach number. The Mach number is the ratio of flow speed to the local sonic speed, or in this case the ratio of flow velocity to the pulse wave velocity. It is also termed the velocity fluctuation ratio (VFR). Recalling that to assume linearity of the arterial system, the flow velocity should be small as compared to the pulse wave velocity, or that VFR should be small. The allometric equation for velocity in the aorta is calculated from the stroke volume (V_s), aortic diameter (D) and heart rate (f_h), noting that the stroke volume is the volume flow (Q) during the cardiac period (T),

$$V_s = QT$$
$$= \left(v\pi D^2/4\right)\left(1/f_h\right) \tag{7.9}$$

or

$$0.66W^{1.05} = v\pi\left(0.205W^{0.36}\right)^2 / \left(3.60W^{-0.27}\right)$$

$$v = 18W^{0.06} \tag{7.10}$$

Thus, the mean blood flow velocities in the aortas of different mammals are about the same and have only a very slight dependence on their body weight.

In terms of allometric relations (Table 7.1), the Mach number can be expressed as:

$$Ma = v/c = 18W^{0.06}/446\,W^{0.0}$$

or

$$Ma = 0.04\,W^{0.06} \tag{7.11}$$

Thus, the velocity fluctuation ratio has a very slight dependence on body weight. In other words, the ratio of mean blood flow velocity to pulse propagation velocity is practically constant i.e., independent of mammalian body weights. Given the density of blood $\rho = 1.06$ g/cm³, viscosity of blood $\eta = 0.03$ P, the aortic diameter, and velocity, the Reynolds number is easily calculated:

$$Re = (1.06)\left(18W^{0.06}\right)\left(2\times0.205W^{0.36}\right)/0.03$$

or

$$Re = 260.76W^{0.42} \tag{7.12}$$

It can be seen that Reynolds number, though dimensionless, is not invariant of mammalian body weights. It is approximately proportional to body length dimensions ($W^{1/3}$). This also illustrates the point that dimensionless Pi-numbers do not necessarily lead to similarity principles.

The requirements for dynamic similarity (Rosen, 1978) are that two flows must possess both geometric and kinematic similarity. Thus the effects of, for instance, viscous forces, pressure forces, and surface tension, (Hossdorf, 1974; Li, 1987) need to be considered. Here we examined only the ratio of inertial forces to viscous forces i.e., Reynolds number and the ratio of inertial forces to compressibility forces, i.e., Mach's number or velocity fluctuation ratio. For a truly incompressible fluid, $c \gg v$ such that $Ma = 0$. For the analysis of blood flow in arteries, both blood and arterial walls are normally assumed to be incompressible. The Poisson ratio (σ_p) for the aorta is about 0.48, close to that of an incompressible material ($\sigma_p = 0.5$, as we discussed

Table 7.2 Calculated Parameters Including Reynolds Number for Four Species of Mammals

	W (kg)	f_h (/min)	D (cm)	L (cm)	v (cm/sec)	Re
Horse	400.0	42.8	3.54	112	25.8	3229
Man	70.0	68.6	1.89	65	23.2	1553
Dog	20.0	96.2	1.21	44	21.5	918
Rat	0.5	260.5	0.32	14	17.3	195

in Chapter 6). The assumptions of linearity and linear system analysis applied to hemodynamic studies often require that the ratio v/c ≪ 1, or that the diameter of the blood vessel is small compared to the pulse propagation wavelength. This is justified during the large part of the cardiac cycle. At peak flow rates in early systole, however, the ratio of v/c is large (but not exceeding 1), and turbulence may ensue to produce nonlinear effects.

Reynolds number, from Equation (7.12), is seen to vary with body length dimensions or the diameter of the aorta. If we were to interpret the effects according to this allometric equation, small mammals, such as a 500-g rat (Re = 195, Table 7.2) are expected to have a turbulence-free aortic flow. Earlier we mentioned that turbulent flow can occur if the Reynolds number exceeds 2000. Now, mammals larger than the man, such as a 400-kg horse, will have a calculated Reynolds number of 3229, far exceeding the critical number of 2000. From this simple calculation, it appears at first glance that the flow in these larger mammals would encounter frequent turbulence in their ascending aortas. Disturbed flow in the flow velocity waveform is normally associated with turbulence. This was observed in peak systole measured with hot-film anemometers in the thoracic aorta, but was not observed in the abdominal aorta of the horse. The value calculated for the ascending aorta of a 70-kg man is about 1553, just below the turbulence threshold level.

One question that immediately arises is whether the resulting Reynolds numbers calculated for large mammals, such as the horse, show that turbulence may occur for a large portion of the systole in the aorta. If this were the case, it would contradict previous observations of the similar pressure and flow waveforms in the aortas of mammals. In fact, this may not necessarily be the case. It has been well documented that turbulence may not exist even for Reynolds numbers greatly exceeding the critical value of 2000. It is only established that for Reynolds numbers under 2000, turbulence does not normally occur. Again, this value was established under steady flow conditions in rigid tubes.

The arterial blood flow exhibits pulsatile characteristics, and peripheral outflow occurs mostly in diastole. In systole during ventricular ejection, the aorta distends as a reservoir to accommodate the flow as described by the classic windkessel model of the arterial system. In concert with the pulsation, this compliance of the aorta acts to protect the peripheral vascular beds from sudden surges in pressure and flow. The compliance, defined as the ratio of change in volume due to a change in pressure,

$$C = dV/dP \qquad (7.13)$$

is allometrically proportional to body weight, as we shall see in the next section, since blood pressure is relatively constant throughout the mammalian species (Table 7.2).

It is also known that turbulence is frequency (or in this case heart rate) dependent. A larger volume change occurs in the aorta of a larger mammal, and the longer effective length of the aorta and a much slower heart rate all help to reduce the tendency of turbulence to reside in too large a portion of systole. Conversely, though a smaller mammal has a lower calculated Reynolds number, the pulsatile volume change is smaller, the effective length is shorter, and the heart rate is much faster, this does not allow it to avoid turbulence altogether. Thus, the compliance of the aorta, the effective entrance length, and the heart rate are also determinants of turbulence. The consequence of such an interplay is the manifestation of similar aortic flow waveforms observed for the mammalian species.

7.2 BLOOD PRESSURE AND FLOW WAVEFORMS

Similar pressure and flow waveforms recorded in mammalian aortas suggest that pulse transmission characteristics may also be similar. I shall elaborate on the validity of this hypothesis in this section utilizing allometric equations of pertinent hemodynamic parameters and the already-familiar windkessel model of the arterial system.

As we have shown previously, the recordings of pressure and flow waveforms in corresponding parts of the arterial systems in some mammalian species are similar, particularly in the aorta. We have learned from Chapter 6 that pressure and flow pulses are modified as they travel away from the heart due to, for instance, wave reflections and damping. In the aorta, the more dominant factor appears to be wave reflection. From Chapter 5 we see that mammalian left ventricles perform a similar mechanical function, which results in an ejection of blood into an arterial load of similar function. We shall perform a similarity analysis on this latter, that is, to utilize allometric equations of relevant hemodynamic variables in order to examine whether pulse transmission characteristics are similar in the arterial systems of different mammals.

The propagation of the pressure pulse in the aorta depends on the geometric and elastic properties of its wall, on the density of the blood contained in it, and on wave reflection. This latter is due to the mismatch of impedances at the level of peripheral vessels. We have shown that the input impedance (Z_{in}) can be used to characterize these global properties of the arterial system. The fraction of the propagating pulse that is reflected is given by, naturally, the reflection coefficient, as shown before:

$$\Gamma = \left(Z_{in} - Z_o\right) / \left(Z_{in} + Z_o\right) \qquad (7.14)$$

where Z_o denotes the characteristic impedance of the proximal aorta. Pulse transmission is associated with a propagation constant γ, as we discussed in Chapter 6,

$$\gamma = \alpha + j\beta \qquad (7.15)$$

where α is the attenuation coefficient, describing pulse damping due to viscous losses, and β, the phase constant, denotes the relative amount of phase shift or time delay.

In the present analysis γ and Γ are computed from allometric equations for selected species of mammals with grossly different body weights.

The fact that the aorta is mostly elastic with little viscous losses allows us to neglect the attenuation component of the propagation and to approximate γ as

$$\gamma \simeq j\beta \qquad (7.16)$$

β can be expressed in terms of the pulse wave velocity, c,

$$\beta = \omega/c \qquad (7.17)$$

where $\omega = 2\pi f_h$, and f_h = heart rate (s^{-1}). The wave velocity c can be represented in its allometric form as

$$c = 446W^0 \text{ cm s}^{-1} \qquad (7.18)$$

and for heart rate,

$$f_h = 3.6W^{-0.27} \text{ s}^{-1} \qquad (7.19)$$

hence

$$\beta = 0.051W^{-0.27} \text{ cm}^{-1} \qquad (7.20)$$

We shall now approximate the input impedance of the systemic arterial tree by the three-element windkessel model of the systemic arterial system. This simplifies the computation. For this representation, the input impedance can be expressed in terms of its elements as

$$Z_{in} = Z_o + R_s/(1 + j\omega CR_s) \qquad (7.21)$$

where, as defined previously, R_s is the systemic peripheral resistance and C is the total systemic arterial compliance. The allometric form of these parameters are

$$R_s = 3.06 \times 10^4 \text{ W}^{-0.68} \text{ dyn s cm}^{-5} \qquad (7.22)$$

$$C = 0.18 \times 10^{-4} \text{ W}^{0.95} \text{ g cm}^4 \text{ s}^2 \qquad (7.23)$$

Thus, the peripheral resistance decreases, while systemic arterial compliance increases with mammalian body size (Li and Noordergraaf, 1991).

Experimental values of Z_o/R_s under normal physiological conditions have been reported to be

$$0.06 < Z_o/R_s < 0.10 \qquad (7.24)$$

We can now calculate the reflection coefficient for each harmonic from Equation (7.14). The magnitudes of the reflection coefficients for the first six harmonics of

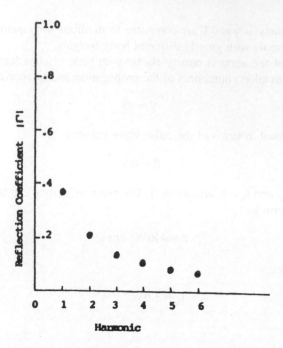

Figure 7.1   Reflection coefficients of the first six harmonic components calculated from allometric equations. The formula $Z_o/R_s = 0.1$ was used. (From Li, J. K-J., and A.
Noordergraaf, *Am. J. Physiol.*, 261:R519–R521, 1991. With permission.)

Table 7.3   Data for Different Mammals

	Body weight W (kg)	Heart rate f_h (/min)	Phase velocity c (cm/sec)	System length L (cm)	Reflection coefficient Γ	Γ(exp)	Propagation constant × L γL (nondimensional)
Horse	400	36	400	110	0.36	0.42	1.13
Man	70	70	500	65	0.38	0.45	1.06
Dog	20	90	400	45	0.39	0.42	1.01
Rabbit	3	210	450	25	0.41	0.48	0.93

Note: $Z_o/R_s = 0.1$ is used in the calculation of Γ and γ.

pressure are plotted in Figure 7.1. This figure illustrates that the amount of reflected
wave is large at low frequencies and much smaller at higher frequencies. This
propagation wavelength dependence has been explained in Section 6.3. The calculated magnitudes of the reflection coefficient for the fundamental harmonic component using $Z_o/R_s = 0.1$ for horse, man, dog, and rabbit are shown in Table 7.3. The
values prove to be independent of body weight.

Equations that are used to describe pulse wave transmission generally contain
exponents of the form γL. For instance, the propagation constant is defined for a
pressure wave propagating without reflection over a distance L as

$$p = p_o\, e^{-\gamma L} \tag{7.25}$$

For the propagation through the entire length of the aorta, L will be the aortic length.
This is expressed in allometric form as (Li, 1987)

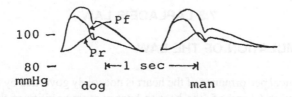

100 —

80 —
mmHg   dog   man

Figure 7.2 Recorded root aortic pressure waveforms in human and dog resolved into their forward (P_f) or antegrade and reflected (P_r) or retrograde components. Note the similarity of the waveforms in spite of body weight and heart rate differences. (From Li, J. K-J., and A. Noordergraaf, *Am. J. Physiol.*, 261:R519–R521, 1991. With permission.)

$$L = 17.5W^{0.31} \qquad (7.26)$$

the value of γL becomes

$$\gamma L = j\beta L$$
$$= j0.893W^{0.04} \qquad (7.27)$$

showing only a very slight dependence on body weight. The calculated values for different mammals are listed in Table 7.3.

Our earlier findings in Chapter 6 have shown that the ratio between the pulse propagation wavelength, λ, for the fundamental harmonic, and the length of the aorta, L, equals about 6, independent of the body weight of the mammal. The results presented here, that γL equals about 1, again independent of the mammalian body weights, confirm the earlier findings that the propagation characteristics are similar. In addition, it was shown here that the global reflection coefficient for the same harmonic is practically invariant. This occurs in spite of the vast differences in parameter values, such as heart rate, systemic peripheral resistance, total systemic arterial compliance, and aortic characteristic impedance, that are associated with different body weights. The hemodynamic allometry-predicted fundamental reflection coefficients compared favorably with those obtained from experimental values (Γ_1(exp), Table 7.3). These latter were calculated by applying the time-domain method (Chapter 6) on the pressure and flow waveforms.

These observed phenomena concerning pulse transmission, pulse wave velocity, and input impedance, as we discussed in Chapter 6, must all be attributed to a common mechanism. The architecture of a branching arterial junction is such that only a portion of the pulse wave generated by the ventricle reaches the capillaries. Another part is reflected by the peripheral vessels in the arteriolar range. Reflected waves encounter mismatched branching sites on their return trip to the ventricle. As a result, a negligible fraction of the pulse wave actually reaches the root of the aorta, with the exception of the fundamental frequency component for which the wavelength is longer than the effective length of the vascular system. Since reflection coefficients as well as the ratio of wavelengths to system length appear independent of body weight, similarity in pressure waveforms resolved into their forward or antegrade and reflected or retrograde components follows, as shown in Figure 7.2.

7.3 LAPLACE'S LAW

7.3.1 FORMULATION OF THE LAW

The mechanical performance of the heart is not solely governed by Starling's law. As we have seen in Chapter 5, the beat-to-beat pumping ability of the mammalian heart is determined by its force-generating capability and the lengths of its constituent muscle fibers. The formula for calculating force or tension, however, has been based on the law of Laplace:

$$T = pr \tag{7.28}$$

It states that the pressure difference, p, across a curved membrane in a state of tension is equal to the tension in the membrane, T, divided by its radius of curvature, r (Woods, 1892). To apply this formula, a certain geometric shape of the heart has to be assumed to arrive at the radius or radii of curvature. The ventricle has therefore been described geometrically as either a thin- or thick-walled sphere or ellipsoid (e.g., Koushanpour and Collings, 1966; Mirsky, 1974; Weber and Hawthorne, 1981) as shown in Figure 7.3.

In the case of an ellipsoid, there are two principal radii of curvature, r_1 and r_2. Laplace's law dictates

$$p = T\left(1/r_1 + 1/r_2\right) \tag{7.29}$$

Pressure, p is actually the transmural pressure, i.e., the pressure on one side of the ventricular wall minus the pressure on the other. Tension, T, is expressed in dynes per centimeter, and r_1 and r_2 in centimeters are the long- and short-axis radii.

For a sphere, $r_1 = r_2$, we have

$$p = 2T/r \tag{7.30}$$

In a cylinder such as a blood vessel, one radius is infinite, so that

$$p = T/r \tag{7.31}$$

which indicates a lesser tension in the wall needed to balance the distending pressure. This is illustrated in Figure 7.4.

Mean arterial blood pressures are practically the same for mammalian species,

$$p = 1.17 \times 10^5 \ W^{0.033} \ dyn/cm^2$$

or

$$p = 87.8 W^{0.033} \ mmHg \tag{7.32}$$

which can be assumed to approximate the mean left ventricular ejection pressure.

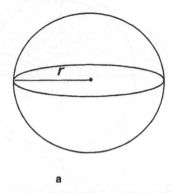

a

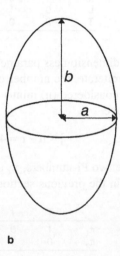

b

Figure 7.3   (a) Spherical and (b) ellipsoidal geometric models of the left ventricle.

Equation (7.28) implies that the larger the size of the heart, the greater the tension exerted on the chamber musculature. This becomes more apparent when Equation (7.32) is considered; i.e., since the distending pressure is constant, tension varies directly with the radius in all mammalian species. To sustain this greater amount of tension, the wall must proportionally thicken with increasing radius of curvature. This results in a larger heart weight as observed in a large mammal.

7.3.2   DIMENSIONAL ANALYSIS

Assume the ventricle can be geometrically represented by a spheroid with a single radius of curvature, r. A dimensional matrix can be formed by first expressing T, r, p, and h in terms of mass (M), length (L), and time (T),

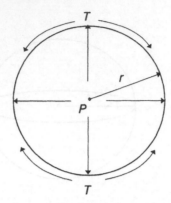

Figure 7.4 Laplace's law relating the distending pressure, radius of curvature to tension.

	T	r	p	h
M	1	0	1	0
L	0	1	−1	1
T	−2	0	−2	0

(7.33)

In order to derive dimensionless parameters (π_i), Buckingham's Pi-theorem will again be utilized. To reiterate, the number of Pi-numbers (j) is equal to the number of physical quantities considered (n) minus the rank (r) of the matrix (Rosen, 1978; Li, 1983). Here we have

$$j = n - r = 4 - 2 = 2 \tag{7.34}$$

Thus, there will be two Pi-numbers, π_1 and π_2. These are obtained by following the procedure shown in the previous section. The results are

	T	r	p	h
π_1	1	0	−1	−1
π_2	0	1	0	−1

(7.35)

The two Pi-numbers expressed in terms of the physical quantities are

$$\pi_1 = T/ph \tag{7.36}$$

and

$$\pi_2 = r/h \tag{7.37}$$

These two equations provide a description of the geometric and mechanical relations of the mammalian hearts. Further, it is clear that Laplace's law is implicit in the ratio of the two, i.e.,

$$I = \pi_1/\pi_2 = T/pr = \text{constant} \tag{7.38}$$

7.3.3 INVARIANT NUMBERS

Since both r and h are proportional to the 1/3 power of body weight, both π_1, and π_2 and their ratio, I, are not only dimensionless; they are also independent of mammalian body weights. This also establishes a scaling factor. They are thus considered invariant numbers. This invariance implies that Laplace's law applies to all mammalian hearts.

7.3.4 COMPARATIVE EXPERIMENTAL EVIDENCE

We have demonstrated that allometric equations are useful when combined with dimensional analysis in the establishment of similarity principles. It needs to be re-emphasized here, however, that in order to utilize Buckingham's theorem, pertinent physical quantities need to be known first, such as T, p, r, and h in the present analysis. The derived Pi-numbers are not necessarily invariant of body weight as in the case of Reynolds number (in the present case, they happen to be invariant), although they are dimensionless.

The derived Pi-number T/pr and r/h need to be used concurrently to represent invariance of the geometric and mechanical aspects of the mammalian hearts. The geometric and anatomic structure of the heart represents its fundamental adaptability and capacity for generating force and external work. Physical deformation, rather than the result of differential growth, was suggested as being the responsible mechanism in the morphological sequence of the development of the heart. However, allometry implies differential growth, according to Huxley.

From a comparative point of view, the applicability of Laplace's equation to mammalian hearts have been examined by Martin and Haines (1970). Assuming an ellipsoidal model of the left ventricle, they wrote

$$c_k = h\left(1/r_1 + 1/r_2\right) \tag{7.39}$$

where c_k is a constant = p/k and tension is directly proportional to wall thickness T = kh. Here, r, and r_1 and r_2 are semiminor and semimajor axes radii respectively. In the seven species of mammals examined with body weight variations of 814-fold, c_k, which expresses wall thickness to radius ratio, a dimensionless constant, was found to be relatively constant. In general, the radius to wall thickness ratios are practically constant and Laplace's law applies to mammalian hearts. These ratios can also be concluded from Li (1983) and Calder (1981).

$$r_1/h = 1.93W^{-0.014}$$

$$r_2/h = 5.41W^{0.008} \tag{7.40}$$

For a 70–kg man, these ratios are 1.82 and 5.6, respectively, or $r_2/r_1 = 3$. In other words, the left ventricular base-to-apex diameter is about three times the short-axis diameter.

Holt et al.'s (1968) data show, by assuming a spherical shape of the heart, the end-systolic (es) and end-diastolic (d) geometries for mammalian left ventricles are

$$V_d = 1.76W^{1.02} \quad \text{hence} \quad r_d = 0.75W^{0.34}$$

$$V_{es} = 0.59V_d^{0.99} \quad r_{es} = 0.627W^{0.34} \tag{7.41}$$

$$= 1.04W^{1.02}$$

where V and r are volume and radius of the ventricle, respectively. These allometric relations conform to the 1/3 power law, as first proposed by Lambert and Teissier (1927). Tension at end-systole is easily calculated, i.e.,

$$T_{es} = pr_{es} = 0.734W^{0.37} \times 10^5 \text{ dyn/cm} \tag{7.42}$$

and for the spherical ventricle,

$$T_{es} = pr_{es}/2 \tag{7.43}$$

This is also directly proportional to the linear dimension of body length ($W^{1/3}$). This result also indicates that the larger the mammalian heart (larger r), the larger proportionally is the tension developed; this explains why ventricular pressure remains practically the same for all mammals.

7.4 THE HEART RATE

Heart rate is the most commonly measured variable in the evaluation of cardiovascular function. As we saw in Chapter 3, Lambert and Teissier (1927) suggested that mammalian heart rates are inversely proportional to body length dimensions. Clark (1927) also reported that the heart rate is inversely proportional to body weight. The relation in its allometric form was reported two decades later by Adolph (1949) as

$$f_h = 3.60W^{-0.27} \tag{7.44}$$

Yet another two decades later, Holt et al. (1968), in the direct measurement of heart rate in nine mammalian species (cattle, horse, swine, sheep, sea lions, goats, dogs, rabbits, and rats), obtained another allometric equation

$$f_h = 236/60 \, W^{-0.25} \tag{7.45}$$

The allometric equation for heart rate is also given by

$$f_h = 4.01W^{-0.25} \tag{7.46}$$

Table 7.4 Allometric Equations of Some Hemodynamic Parameters

Heart rate f_h (sec^{-1})	4.02	−.25	Stahl (1967)
Stroke volume V_s (ml)	0.66	1.05	Holt et al. (1968)
Arterial pressure p (dyn/cm^2)	1.17×10^5	0.0033	Gunther and Guerra (1955)
Metabolic rate MR (erg/sec)	3.41×10^7	0.734	Kleiber (1961)
Heart weight W_h	0.0066	0.98	Adolph (1949)
Oxygen uptake $\dot{V}O_2$ (ml)	11.6	0.76	

This relation, reported by Stahl (1967; Table 7.4) and by the previous investigators, deviates from the true 1/3 power of body weight as suggested by Lambert and Teissier (1927). Thus, a 2000-kg elephant will have a calculated heart rate of 36 beats per minute, whereas the heart of a 3-kg rabbit will beat five times faster at 183 beats per minute despite the 667-fold difference in body weight and equivalent difference in heart weight, which is given by

$$W_h = 0.0058W^{0.98} \tag{7.47}$$

Smaller hearts beat faster. This fact tends to lead one to think that a smaller heart actually requires more energy in a given amount of time per minute for pumping blood to perfuse the vascular beds. Indeed if we take a look at the size of the pump and the number of beats, it is roughly proportional to cardiac output, and also the metabolic rate (Table 7.4),

$$CO = 0.187W^{0.81} \text{ l/min} \tag{7.48}$$

$$MR = 3.41 \times 10^7 \, W^{0.734} \text{ erg/s} \tag{7.49}$$

The relation of heart rate and metabolic turnover rate will be examined in Section 7.5 and in Chapter 9.

Now we shall discuss the relative constancy of the ratio of cardiac output to metabolic rate. Larger mammals have longer aortas and arteries to perfuse a larger cross-sectional area of vascular beds. Consequently, a considerably longer time is required for the blood to reach these vascular beds since the blood flow velocity and pulse wave velocity are relatively constant among mammalian species.

For very small mammals, there is a constraint on the size of the heart and the heart rate it can develop. Cardiac output is the product of stroke volume and heart rate. It appears that the body surface area is of greater importance than body weight in these very small mammals, since they need to maintain "thermoneutrality" or the body temperature in relation to environment temperature.

Thus, in very small mammals, such as the shrew, oxygen consumption and heart size are considerably larger than the expected values and the heart rate is much smaller than expected. This may be because of the conduction velocity, filling, ejection, and relaxation periods required for the cardiac cycle.

We should also note here that there are differences in the excitation pathways among mammalian hearts despite gross similarities in the conduction pattern. This latter is elicited at the pacemaking sino-atrial node, and the propagation of the potential goes through Bachman's bundle to arrive at the atrio-ventricular node, where conduction velocity slows before proceeding to the His bundle, the left and right bundle branches and the purkinje fibers. The spread of excitation in the left ventricle is from the apex toward the base. This is where some of the differences exist.

7.5 ENERGETICS AND EFFICIENCY OF THE MAMMALIAN HEART

The hearts of the mammalian species vary in size, each in a constant proportion to its body weight. For the heart to function as a muscular pump, it requires energy. The energy requirement of its constituent muscle fibers and its ability to generate useful work are of considerable interest (e.g., Starling and Visscher, 1926; Robard et al., 1959; Li, 1983b). They define the mechanical efficiency of the cardiac pump.

In hemodynamic terms, the efficiency of the heart is defined as the ratio of external mechanical work to myocardial oxygen consumption. The external work (EW) is the integral of the product of instantaneous pressure and instantaneous flow, i.e.,

$$EW = \int p(t)Q(t)\, dt \qquad (7.50)$$

Since the pulsatile energy component is generally small compared to the energy required to overcome vascular resistances, the external work can be reduced to

$$EW = \bar{p}V_s \qquad (7.51)$$

where $\bar{p}$ is the mean arterial pressure and V_s is the stroke volume, as discussed before. Thus, EW is also termed stroke work, i.e., the amount of mechanical work performed per heart beat. This again is seen from the P-V diagram (Figure 7.5). The arterial pressure is practically constant throughout the mammalian species, and the stroke volume is proportional to body weight. It was shown previously that the external mechanical work generated by the heart per unit of body weight (or heart weight, for that matter) is constant for all mammalian species. The notion that a heart that can generate a larger stroke work is more efficient is misleading. This leaves us a particular interest in examining whether the efficiency of the heart is constant for all mammals.

Previously, it was shown that the external cardiac work is directly proportional to mammalian body weights (Li, 1983), i.e.,

$$EW/W = constant = \bar{p}V_s/W \qquad (7.52)$$

$$V_s = 0.74W^{1.03} \qquad (7.53)$$

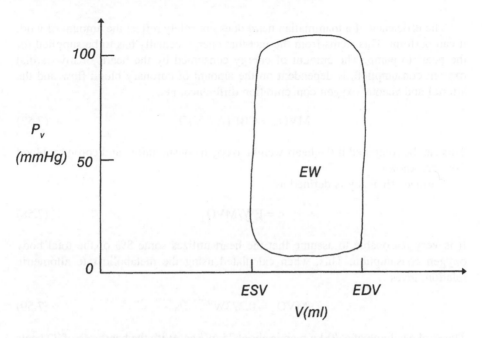

Figure 7.5 Pressure-volume relationship showing the external mechanical work of the ventricle as the area under the loop.

Hence, the external work is related to the generation of tension, taking into account Laplace's law,

$$EW = TV_s/r \qquad (7.54)$$

External work plays an important role in assessing the energy requirements and pumping performance of the heart. Work is defined as force times distance or pressure times volume in the case of an enclosed chamber such as the heart. Along with an increase in coronary blood flow, more myocardial oxygen available to the cardiac muscle will result in an increase in the external mechanical work generated by the ventricle.

The amount of energy liberated by the body is a function of the external work, heat, and energy storage:

$$\text{Energy output} = EW + \text{heat} + \text{energy storage} \qquad (7.55)$$

The amount of energy liberated per unit of time is the metabolic rate. In relation to body weight, Brody (1945) and Kleiber (1947, 1961) determined metabolic rates for both mammals and avians:

$$MR = 70W^{0.75} \qquad (7.56)$$

Normalization of this metabolic rate with body weight can be shown to be directly proportional to heart rate, as will be shown in Section 7.7.

The efficiency of a mammalian heart does not solely reflect the amount of work it can perform. This stems from the fact that energy actually has to be supplied for the heart to pump. The amount of energy consumed by the heart, or myocardial oxygen consumption, is dependent on the amount of coronary blood flow and the arterial and venous oxygen concentration difference, i.e.,

$$MVO_2 = CBF(A - V)O_2 \tag{7.57}$$

This can be computed if the heart weight, oxygen consumption, and coronary blood flow are known.

Cardiac efficiency is defined as

$$e = EW/MVO_2 \tag{7.58}$$

It is very reasonable to assume that the heart utilizes some 8% of the total body oxygen consumption. This, when calculated using the metabolic rate allometric relation, gives

$$MVO_2 = 0.273W^{0.734} \; J/s \tag{7.59}$$

The stroke volume of a 70-kg man is about 75 ml/sec. With the heart rate of 72 beats per minute and a mean pressure of, say, 100 mmHg, we have for each heart beat,

$$EW = 1 \; J/s \tag{7.60}$$

Using allometric equation for calculation of the myocardial oxygen consumption per beat gives a value of 5.14 J/sec. The cardiac efficiency is about 20%. This is a very reasonable estimate,

$$e = EW/MVO_2 = 1/5.14 \approx 20\% \tag{7.61}$$

whereas the mechanical efficiency alone is significantly higher.

7.6 ARTERIOLES, VENULES, CAPILLARIES, AND THE RED BLOOD CELLS

7.6.1   PRESSURE AND FLOW PULSATIONS IN ARTERIOLES AND IN CAPILLARIES

For many decades it has been assumed that flow in the microcirculation, particularly in the arterioles and capillaries, is entirely steady flow. Consequently, Poiseuille's formula has been applied, as before,

$$Q = \left(\pi r^4/8\eta l\right)\Delta p \tag{7.62}$$

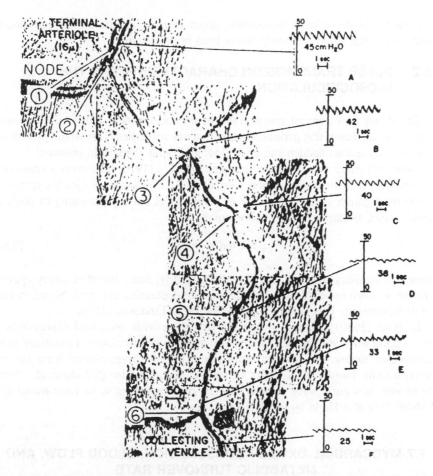

Figure 7.6 Microcirculatory pressure recordings (From Zweifach, B. W., *Circ. Res.*, 34:858–866, 1974. With permission.)

where p is the pressure gradient or pressure drop across a vessel length, l. Steady flow was assumed because of the belief that small peripheral vessels are resistance vessels, thus preventing pulsations from occurring.

It is now known that pulsatile ejection by the ventricule requires only about 10% additional energy for the same stroke volume compared to constant outflow. This minimal additional energy associated with pulsatile ventricular ejection reflects the compliant properties of the receiving arterial tree and facilitates vascular exchanges back and forth.

An appreciable fraction of the energy in the pressure and flow pulses generated by the heart reaches the capillaries in pulsatile form. This can be observed from Figure 7.6. Experimental measurements made by, for instance, Intaglietta et al. (1970) in cat omentum and by Zweifach (1974). This is shown in Figure 7.6, from the direct recording of pressure. It also shows that in the terminal arteriole, the pulse pressure amplitude remains appreciable, about 15 mmHg, with a mean pressure of about 60 mmHg, some 30 mmHg lower than in the arteries.

7.6.2 PULSE TRANSMISSION CHARACTERISTICS IN THE MICROCIRCULATION

The steady flow concept assumed for the microcirculation is in accordance with the windkessel theory that peripheral vessels act as stiff tubes. This would protect the small vessels against sudden surges in flow and rapid changes in pressure.

Estimated phase velocity from Intaglietta et al.'s (1970) data gives a value of 7 to 10 cm/sec, in the microcirculation. Caro et al. (1978) gave an analytical expression for the propagation speed in the case of a sinusoidal pulse propagating through an elastic vessel, assuming blood is Newtonian:

$$c = 1/4 \ d(\omega/\eta D)^{1/2} \tag{7.63}$$

where d is capillary diameter and D is distensibility. Since blood viscosity appears to decrease when measured in capillary tubes of decreasing diameter, blood, in fact, is non-Newtonian. This is known as the Fahraeus-Lindqvist effect.

Li et al. (1980) apparently were the first to provide analytical expressions to predict pulse wave velocity and attenuation in the microcirculation. Linearized pulse transmission theory was utilized. Subsequently, pulse transmission from the left ventricle to the human index finger vessel was accounted for (Salotto et al., 1986). The results show pulse wave velocity of a few centimeters per second and attenuation of about 30% at 1 Hz in large arterioles.

7.7 MYOCARDIAL OXYGEN CONSUMPTION, BLOOD FLOW, AND METABOLIC TURNOVER RATE

The proper performance of physiological functions requires a certain amount of energy. Indeed, this aspect has been the subject of many studies concerning animal body metabolism. A well-known allometric equation relating metabolic rate (MR) to body weight of both mammals and avians was established, as noted before, by Kleiber (1947),

$$MR = 70W^{0.75} \tag{7.64}$$

where MR is in kilocalories per day and W is in kilograms. The metabolic rates calculated for three mammalian species are shown in Table 7.5. The larger mammal is seen to have a larger metabolic rate. The exponent has been reported elsewhere as 0.734 and 0.79, and thus is not directly proportional to body surface area.

In a more recent paper, Kleiber (1975) suggested that metabolic rate per unit body weight should be termed the metabolic turnover rate (MTR), i.e., the fractional rate of metabolism converted from the chemical content of the body. Accordingly, the following results:

Table 7.5 Physiological Parameters of Three Mammals

	Body weight W (kg)	Metabolic rate MR (J/sec)	Metabolic turnover rate MTR (J/sec/kg)	Heart rate f_h(min^{-1})	(J/kg)	(EW/W)/(MTR/f_h) dimensionless
Man	70	82	1.17	70	1.0	0.014
Dog	20	32	1.6	90	1.07	0.013
Guinea pig	0.7	2.6	3.7	240	0.93	0.015

$$MR/W = MTR = 70W^{-0.25} \tag{7.65}$$

MTR thus represents the metabolic rate normalized with body weight. As pointed out by Kleiber from data presented by Schmidt-Nielsen (1970), this normalization showed a tremendous increase in the normalized metabolic rate of very small mammals. These can be illustrated in Table 7.5, where MTR is calculated for the same three mammals. Now, instead of the man, the guinea pig appears to have a much higher "normalized" metabolic rate. This led Kleiber to further discuss the limited usefulness of the metabolic turnover rate. I shall illustrate below that the physiological meaning of metabolic turnover rate is useful and significant when considering functional design features of the mammalian cardiovascular system.

In mammals, the heart occupies about 0.6% of body weight. Gunther and De La Barra (1966) quoted an allometric formula of Adolph (1949) of

$$W_h = 6.6 \times 10^{-3} \, W^{0.98} \tag{7.66}$$

where the heart weight W_h and W are both in grams. As shown before, Holt et al. (1968) gave

$$W_h = 2.61W^{1.10} \tag{7.67}$$

with W_h in grams and W in kilograms. In man and dogs, the left ventricle weight is about 60% of the heart weight (Holt et al., 1968).

We have seen that

$$EW \propto W_h \propto W \tag{7.68}$$

This indicates that a larger mammal would provide a greater amount of external ventricular work, in accordance with the larger metabolism it generates.

Cardiac output, a common clinical index of ventricular functional state, is a product of the stroke volume and the heart rate (f_h),

$$CO = V_s f_h \tag{7.69}$$

The exponent for f_h has been reported to be 0.25 to 0.27 for mammalian hearts. Thus, cardiac output is proportional to the three-quarter power of mammalian body weight:

$$CO \propto W^{0.75} \propto MR \tag{7.70}$$

Cardiac ouput is now a direct function of metabolic rate, as one would expect when examining Kleiber's equation (7.64; White et al., 1968).

Normalization of these hemodynamic variables with body weight would generate the following interesting results:

$$EW/W = \text{constant} \qquad\qquad (7.71)$$

$$CO/W \propto W^{-0.25} \propto f_h \propto MTR \qquad\qquad (7.72)$$

These results are also of considerable physiological importance. The ventricular external work per unit body weight of Equation (7.71), or per unit of heart weight, termed the cardiac external work intensity, is invariant among mammals. For man, taking $V_s = 75$ ml, $p = 100$ mmHg, $W = 70$ kg, and $W_h = 370$ g, the external work is about 1 J and the constant is about 2.7 J/kg of heart weight. That is, per unit of heart weight,

$$EW/W_h = 2.7 \text{ J/kg}$$

or

$$EW/W = 1/70 \text{ J/kg} \qquad\qquad (7.73)$$

per unit of body weight, for all mammalian hearts. EW, we recall, is the left ventricular external work. For a 2100-kg elephant, one would expect its left ventricular external work to be 30 J. For a 280-g rat, this value would be just 4 mJ.

The second result, Equation (7.72), relates metabolic turnover rate to heart rate. In other words,

$$MTR/f_h = \text{constant} \qquad\qquad (7.74)$$

This constancy is expected to exist for the selected three species of man, dogs, and guinea pigs, as calculated in Table 7.5. This means that for any given mammal, the metabolic turnover rate during each cardiac cycle is a constant. For a 70-kg man, this constant is about 0.24 cal/kg, assuming a heart rate of 70 min. Introducing the conversion, 1 cal = 4.18 J, gives

$$MTR/f_h = 1.0 \text{ J/kg} \qquad\qquad (7.75)$$

This result readily explains the large normalized metabolic rate, of very small mammals observed by Schmidt-Nielsen (1970) and disagreed upon by Kleiber (1975). Smaller mammals have much faster heart rates, which are associated with greater metabolic turnover rates.

7.7.1   ANOTHER NEW SIMILARITY PRINCIPLE

Dimensional analysis has been the basis for establishing physical laws and for deriving physical constants. For biological systems, a similarity principle is established if it is independent of mammalian body weight and is also dimensionless. Examination of the dimensions of Equations (7.71) and (7.74) gives

$$EW = k_1[M][L][T] \tag{7.76}$$

and

$$MTR = k_2[M][L][T] \tag{7.77}$$

where [M], [L], and [T] are mass, length, and time, and k_1, k_2 are numerical constants. The dimensionless similarity criterion is thus

$$(EW/W)/(MTR/f_h) = k_1/k_2 \tag{7.78}$$

or that

$$[M]^0 [L]^0 [T]^0 = \text{dimensionless constant} \tag{7.79}$$

Stated in words: the external ventricular work intensity during each cardiac cycle is directly proportional to the metabolic turnover rate. This constant of proportionality is an invariant among mammals. Equation (7.78) gives the "cardiac work-rate number".

CHAPTER 8

Closed-Loop Analysis of the Circulation

8.1 FUNDAMENTALS OF BIOLOGICAL CONTROLS

8.1.1 DEFINITIONS

A biological control system consists of natural physical components that are related to form an entity in order to perform the functions of regulation, command, and direction of the physiological system.

An example of a biological control system is the blood pressure control, where the blood pressure in a mammalian species is regulated in such a way that it stays constant within a narrow range of a mean pressure of about 100 mmHg. Body temperature control is another example.

For any control system, there are input and output. The input is the stimulus or excitation applied to a biological control system from an external energy source in order to effect a specified response from the control system. The output is the actual response obtained from a control system. With the knowledge of the output to a specific input, it is possible to identify the nature or characteristics of the system's component. For a given input to a control system, the output can also be readily obtained. Conversely, if the output and the system characteristics are known, the input can be identified.

A biological control system may have more than one input or output. The system can be either open or closed loop. In an open-loop system, the control action is independent of the output. In a closed-loop system, the resulting control is dependent on the desired output.

An open-loop system is usually faced with instability, although with simpler control. Most biological systems, however, are closed-loop or feedback control systems; these systems can achieve the desired response by feeding back the output to be compared with the input to the system so that appropriate control action may be formed as functions of both the input and the output.

Figure 8.1 illustrates an open-loop system, with the system represented by a block diagram and with an input entering the system and the resulting response as an output. The system may perform certain functions, such as rate-sensitive changes (d/dt).

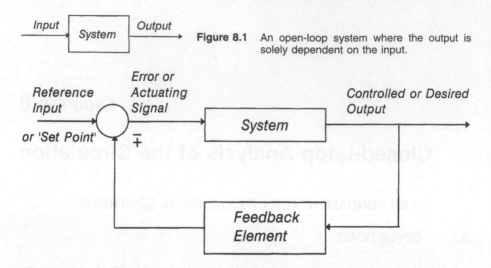

Figure 8.1   An open-loop system where the output is solely dependent on the input.

Figure 8.2   A closed-loop control system where the output can be modified by adjustment of the error signal, which compares the desired output with the input.

For a closed-loop system, the output may be added to the input, as in the case of positive feedback, or it may be subtracted to the input, as in the case of negative feedback. As an example, the nerve action potential generation is through positive feedback, since the potential can add to achieve the threshold and can go beyond threshold. Blood pressure and body temperature controls, however, are negative feedback control systems. Figure 8.2 illustrates the closed-loop control system, with a summing point. The error signal is obtained at this point by either adding (positive feedback) or subtracting (negative feedback) the output from the input. As can be seen, a negative feedback system can achieve stability simply because the error from the desired output is progressively reduced, or "minimized".

Any control system, especially in a biological system, can have disturbances or noises entering it. In addition, there may be several control elements within the system with variable "gain". A summary of a more complex representation is shown in Figure 8.3.

8.2 ALLOMETRY IN CARDIOVASCULAR CONTROL

We shall make a brief introduction to this new field of analysis. We shall begin by illustrating the open-loop control relationship. This is best served by looking at the response to the operation of Starling's law of the heart. The end-diastolic volume of the proceeding beat dictates the stroke volume of the subsequent beat, with other parameters remaining constant. This is simply shown in Figure 8.4. Knowing the allometric relations of the EDV and the SV, we have a relation known as the ejection fraction (EF), which determines the mammlian cardiac function if all other factors remain constant. The ejection fraction is shown by the block representing the ventricle.

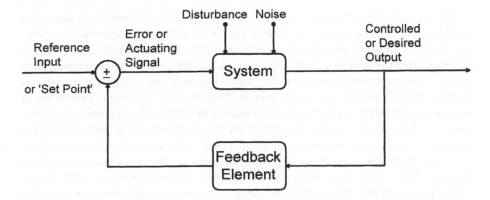

Figure 8.3 A closed-loop control system with disturbance and noise inputs.

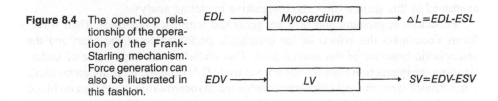

Figure 8.4 The open-loop relationship of the operation of the Frank-Starling mechanism. Force generation can also be illustrated in this fashion.

If we consider other factors, we can then formulate a closed-loop control system, where the stroke volume is subjected to modification by the afterload due to the arterial system. In this case, we know that the higher the peripheral resistance and the lower the arterial compliance, the lower the stroke volume, and the inverse is true.

The arterial system can be represented, as we have shown, by a lumped three-element windkessel model. Thus, R_s, Z_o, and C can all alter the stroke volume within the beat. The resulting changes are translated to the venous return to the heart, which then modifies the end-diastolic volume. This simplified view suffices to illustrate the closed-loop relationship of the hemodynamic function.

We can, of course, perform a much more elaborate illustration by considering blood pressure control, heart rate, the coronary circulation, and myocardial contractility changes. Thus, the "black box" representing the heart is now subjected to parametric variations. A dimensional matrix can be used to represent the properties of the left ventricle. One can logically then obtain the Pi-numbers that govern the function of the heart (such as those illustrated in Chapter 7). One can also include the venous and the microcirculatory systems.

8.3 HEART-ARTERIAL SYSTEM INTERACTION

In this section, I will deal with the function of the heart and the way in which the arterial system can modify its response through a closed-loop adjustment.

The pumping heart and the receiving arterial system are such close functional complements that the circulatory system cannot be effectively described by either one. Only by virture of the arterial system can oxygen, humoral agents, and nutrients be transported to the vital organs of the body, while the heart provides the necessary energy. The pulsation observed in the arterial system and the energy transfer is closely coupled, as we saw in Chapter 6 (Section 6.5). The force generation in different mammalian hearts has been examined in Chapter 7. According to a new similarity principle for cardiac energetics, the external work of the left ventricle per unit body weight is a constant, and that metabolic rate per unit of body weight is directly proportional to heart rate (Li, 1983). Allometric relation of body metabolism (Kleiber, 1961, 1975) has been a subject of debate (Schmidt-Nielsen, 1975; Yates, 1981; Heusner, 1982, 1983, 1984). In addition, the relationship between myocardial oxygen consumption, external mechanical work, cardiac efficiency, and body metabolism has not been analyzed. These will be discussed in Chapter 9. The similarity of functional interaction of the heart and the arterial system in mammals will be examined in this section through comparative modeling analysis.

Pulsations in pressure and flow are generated with each heartbeat. Their waveforms encompass the effects of the contractile performance of the heart and the viscoelastic behavior of the arterial load. This underlies the importance of understanding the interaction between the left ventricle (LV) and the arterial system (AS). Furthermore, the coronary vasculature within the myocardium is dependent on blood pressure in the aorta, which serves as its perfusion pressure. This forms a natural closed-loop system. Coronary blood flow occurs mostly in diastole when the aortic valve is closed. This indicates that the effects of arterial wave reflections play an important role in the interaction. Compliance is a dominating component in pulsations. For these reasons, I shall choose to analyze the dynamics of LV and AS interaction in terms of arterial compliance and wave reflections.

8.3.1   THEORETICAL ANALYSIS AND EXPERIMENTAL DATA

In considering the dynamics of LV-AS interaction, it is worth noting that the windkessel model does not afford pulse-transmission characteristics. In addition, it treats compliance as a constant quantity. It is well known, however, that arteries stiffen when pressurized. This means that compliance is a function of pressure. This model is modified to include the nonlinear pressure-dependent compliance (Li et al., 1990; Li, 1993; Li and Zhu, 1994) and will be referred to as the "Li model" with

$$C(p) = a \exp(bp) \tag{8.1}$$

or

$$E_a(t) = 1/C(p) \tag{8.2}$$

where a and b are empirical constants determined from least-squares fitting of simultaneously measured aortic pressure and diameter data. This exponential repre-

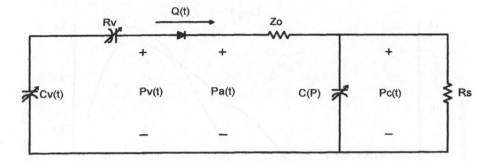

Figure 8.5 Analog model of the left ventricle (LV) and the arterial system (AS) with LV represented by a time-varying compliance and a resistance R_v, and with the AS represented by the Li model with a pressure-dependent compliance C(p), characteristic impedance Z_o and resistance R_s.

sentation has a correlation coefficient of r > 0.97. This Li model coupled to the LV model is shown in Figure 8.5.

The nonlinear aortic compliance can be obtained from the knowledge of aortic pressure and flow (Li et al., 1990). Compliance decreases with increasing pressure. The exponent b therefore has a negative value.

Pulse-transmission characteristics can be examined in terms of wave propagation and reflections, as we have learned before. Pressure (p) and flow (Q) waveforms in the aorta can be resolved into their forward and reflected waveforms, as shown in Chapter 6,

$$p_f = (p + QZ_o)/2 \qquad (8.3)$$

$$p_r = (p - QZ_o)/2 \qquad (8.4)$$

The left ventricular elastance E(t) is defined by the measured left ventricular pressure (p_v), and dead volume (V_d),

$$p_v(t) = E(t)[V(t) - V_d] - R_v Q(t) \qquad (8.5)$$

and

$$V(t) = EDV - \int Q(t) \, dt \qquad (8.6)$$

assuming the flow to the coronary is small during ejection; EDV is the end-diastolic volume. E(t) is readily computed by assuming values for the ejection fraction (0.5 to 0.7) and V_d for a particular mammalian species.

Figure 8.6 illustrates the pressure dependence of aortic compliance where the arterial elastance is computed as 1/C(p) and plotted for the systole. The left ventricular elastance is also computed for the same beat to show the temporal relationship

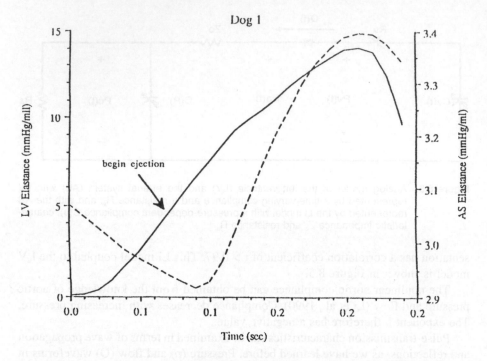

Figure 8.6 Left ventricular and arterial elastances as a function of time during systole.

between the two elastances. Notice that arterial elastance reaches the maximum at about end-systole. Maximum elastance of the left ventricle or E_{max} occurs at end-systole. This reflects the "compliance matching" characteristics of the heart and the arterial system.

Time-domain-resolved forward and reflected waves in the aorta are shown in Figures 8.7 and 8.8 for control and descending thoracic aorta (DTA) occlusion conditions, respectively. During DTA occlusion, the reflected wave arrives earlier in systole and is greater in amplitude, particularly in mid to late systole and in diastole. These reflected waves returning earlier in systole can exert profound influence on ventricular function.

Wave reflections arise in the arterial system, whether systemic (Li et al., 1984) or coronary due to impedance mismatching. The differences in impedances in the arterial tree occur as a result of variations in elasticity and geometry.

Elastic nonuniformities in the arterial tree range from a compliant, less viscous aorta to the stiffer, more viscous peripheral vessels, including the coronaries. The variation is generally attributed to the differential contents of elastin, collagen fibers, and smooth muscle cells within their respective walls. The extent of arterial disten-sion is dependent on the pulsating pressure. The pressure dependence of aortic compliance is well known; compliance in general decreases with increasing pressure (Zhu, 1993; Li and Zhu, 1994). Its variation in the cardiac cycle is demonstrated here. The minimum compliance occurs at about end systole, when the left ventricular elastance is at its maximum, or E_{max}. It is important to differentiate the dynamic

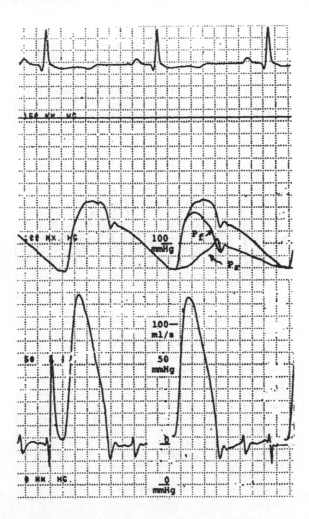

Figure 8.7 ECG, aortic pressure, and flow measured during control. Resolved forward (P_f) and reflected (P_r) waves are also shown. (From Li, J. K-J., *Angiol. J. Vas. Dis.*, 40:730–735, 1989. With permission.)

elastance interaction (Figure 8.6) from steady-state analysis, which considers only mean values.

The pressure-volume relationship well known in the study of left ventricular function has been extended for LV-AS interaction studies. The maximum elastance, or the slope of the end-systolic (ES) P-V relation is based on a time-varying compliance model of the LV (Sagawa, 1978). This has been analyzed by some investigators with E_a, an effective elastance of the arterial system, a nonphysiological quantity that lumps together the steady-state AS characteristics based on the three-element windkessel model. Matching of E_{max} and E_a was evaluated under varied physiological conditions. It should be noted that elastance is a global functional

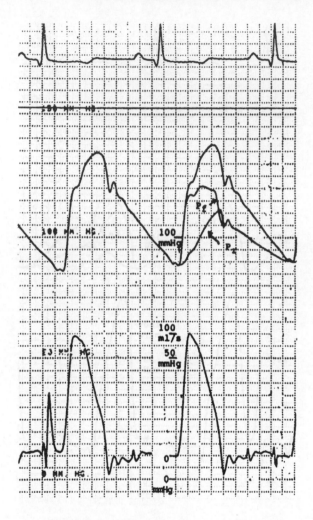

Figure 8.8 Same as in Figure 8.2, during aortic occlusion. (From Li, J. K.-J., *Angiol. J. Vas. Dis.*, 40:730–735, 1989. With permission.)

parameter, whereas elasticity is a parameter that represents the properties of the wall materials. Since

$$E_{max} = ESP/(ESV - V_o) \qquad (8.7)$$

and

$$E_a = R_s/T \qquad (8.8)$$

where V_o is the dead volume at zero left ventricular pressure, this approach investi-

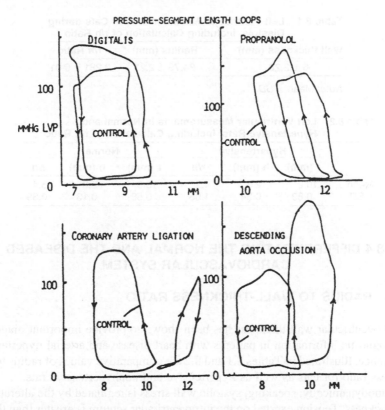

Figure 8.9 Left ventricular pressure-segment length loops. Interventions are compared to their respective controls.

gated only steady-state matching characteristics. Additionally, E_{max} varies greatly with mammalian body weights (hence, heart weights), larger in smaller mammals.

The arterial system can exert a profound effect on regional myocardial function, as seen from cardiac muscle segment shortening measurements during acute (10 sec) descending thoracic occlusion (pressure overload), left anterior descending (LAD) coronary artery ligation (1 hr), and beta adrenergic blockade with propranolol (1 mg/ kg, intravenously) infusion. Descending aortic occlusion increased segment work as assessed by pressure-segment loop area, but decreased shortening. LAD occlusion resulted in a negative value, because of passive lengthening. Propranolol decreased shortening slightly, but shifted the loop rightward, suggesting a depressed cardiac function, although segment work was somewhat reduced (Figure 8.9). In the case of coronary arterial disease, the depressed cardiac performance is also associated with a reduced compliance of the aorta, increased characteristic impedance, and peripheral resistance.

Table 8.1   Left Ventricular Measurements in Cats during
Diastole, Including Calculation of r/h Ratio

Wall thickness (mm)	Radius (mm)	r/h Ratio
6 ± 2.28	24.79 ± 2.21	0.281 ± 0.01

Note: Mean ± *SD.*

Table 8.2   Left Ventricular Measurements in Normal and
Hypertensive Rats, Including Calculation of r/h Ratio

	Hypertensive			Normal		
	r (mm)	h (mm)	r/h	r (mm)	h (mm)	r/h
Mean	10.2	2.5	4.2	10.9	2.0	5.4
± SD	0.89	0.34	1.05	0.58	0.13	0.39

8.4 DIFFERENTIATING THE NORMAL AND THE DISEASED CARDIOVASCULAR SYSTEM

8.4.1   RADIUS TO WALL-THICKNESS RATIO

Left ventricular wall thickness has been shown to provide important diagnostic and prognostic information in patients with heart disease and arterial hypertension, for instance. Illustrated in Tables 8.1 and 8.2 are comparative values of radius to wall thickness ratio for cats as well as hypertensive and nonhypertensive rats.

Hemodynamically, speaking, systolic wall stress is regulated by the alteration in wall thickness. Tension exerted on the intraventricular septum is greater than that on the free wall due to the larger radius of curvature. This indicates the deviation from the assumption of the spherical geometry for the left ventricle. Despite the complex shape of the myocardium, however, the average wall stress is maintained constant throughout. Compensatory increments in muscle mass that accompany an excessive hemodynamic load (e.g., pressure overload) serve to normalize both wall stress and myocardial energetics. This can be seen in myocardial hypertrophy, which occurs in many forms of cardiac disease. Consequently, an increase in wall thickness is observed.

Interest in geometric alterations of the diseased ventricle has recently grown. It has been shown that the slope of the radius of curvature to wall thickness (r/h) is a geometric constant (in dogs, cats, rats, and humans) that determines the mural force at any given transmural pressure. Studies in patients have shown that the r/h ratio can distinguish between reversible and irreversible impairment of myocardial performance. In addition, r/h ratio is found to be relatively constant throughout the ejection and diastolic phases. It is also linearly correlated to left ventricular strain. In coronary artery occlusion, segmental lengthening and corresponding wall thinning have been observed. In the case of hypertrophy, wall thickness increases, as discussed above, returning the r/h ratio toward normal (Li, 1986).

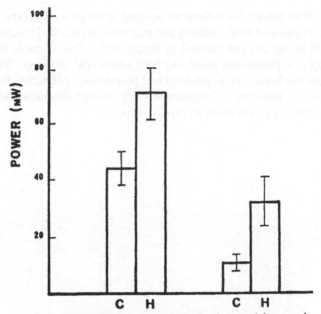

Pulsatile powers of the fundamental harmonic associated with the forward (left) and reflected (right) waves during control (c) and acute hypertension (H).

Figure 8.10 Pulsatile powers of the fundamental harmonic associated with the forward (left) and reflected (right) waves during control (C) and acute hypertension *H.* (From Li, J. K.-J., *Angiol.*, 40:730–735, 1989. With permission.)

8.4.2 THE ROLE OF WAVE REFLECTIONS

Increased afterload is normally associated with hypertension; consequently, the external mechanical work of the heart increases to overcome this load. The fundamental harmonic power of the forward wave is also increased, associated with an increase in reflected wave. As a consequence, the effective power or energy available for transmission is somewhat reduced (Figure 8.10). Increased pulse wave reflections compound the increase in arterial pressure, and the increased work output by the left ventricle is compromised by the reduced efficiency of pulsatile energy transmission to organ vascular beds. In the case of chronic hypertension, an increased arterial pressure load promotes the hypertrophy of the LV.

Wave reflections are altered to a great extent when the arterioles are dilated due to exercise, the strain phase of the Valsalva maneuver, and vasodilators. The peripheral resistance as well as systolic and diastolic pressures decrease under these conditions, and peripheral beds increase in total cross-sectional area, reducing the impedance to flow and improving the impedance matching. Vasodilation appears to decrease as well as delay wave reflections arriving at the proximal aorta. This results in significantly increased stroke volume and improved myocardial shortening as well as a reduction in the oxygen demand to supply ratio. This is particularly true in current therapeutic procedures involving beta adrenergic blockers. We have also found that beta blockers, such as pindolol and propranolol, primarily improve myocardial function by reducing left ventricular work through decreasing afterload and by improving the oxygen demand to supply ratio.

CHAPTER 9

Optimality and Similarity

The close relationship and differences between optimality and similarity will be discussed in this chapter. Understanding that two optimal situations must be dynamically similar, a similarity principle established for mammalian cardiovascular system must necessarily reveal certain dynamic design features. We shall illustrate these by selectively examining the aspect of the energetics of the heart (Section 9.1) and pulse transmission at the branching junctions of the arterial system (Sections 9.2 and 9.3).

9.1 EXTERNAL WORK, OPTIMAL POWER, AND EFFICIENCY

The hearts of the mammalian species vary in size, each in a constant proportion to its body weight. For the heart to function as a muscular pump, it requires energy. The energy requirement of its constituent muscle fibers and its ability to perform useful work are of considerable interest. They define the mechanical efficiency of the cardiac pump.

In hemodynamic terms, the efficiency of the heart is defined as the ratio of external mechanical work to myocardial oxygen consumption. It was shown in Chapter 7 that the external mechanical work generated by the heart per unit of body weight, or heart weight for that matter, is constant for all mammalian species. The notion that a heart that can generate a larger stroke work is more efficient is misleading. It is therefore of particular interest to examine whether the efficiency of the heart is constant for all mammals.

We have seen that the combined use of dimensional analysis, allometry, and hemodynamic principles has proven to be powerful in the establishment of similarity principles. It was established that

$$EW/W = k \qquad (9.1)$$

The fact is that the degree of efficiency of a mammalian heart is not solely reflected in the amount of work it can perform. This stems from the fact that energy actually has to be supplied for the heart to pump. The amount of energy consumed by the heart, or the myocardial oxygen consumption is dependent on the amount of coronary blood flow and the arterial and venous oxygen concentration difference, i.e.,

$$MVO_2 = CBF(A - V)O_2 \qquad (9.2)$$

This can be computed if the heart weight, oxygen consumption, and coronary blood flow are known. Schmidt-Nielsen estimated the arteriovenous O_2 difference of between 4 to 7 ml of O_2 per 100 ml of blood for mammals with 10,000-fold differences in body weight. The cardiac efficiency defined as

$$e = EW/MVO_2 \qquad (9.3)$$

is about 20%, as we have shown in Section 7.5.

Considering Equation (9.1), since heart weight is a constant proportion of the total body weight, then

$$EW/W_h = constant \qquad (9.4)$$

It is important to note that both the external work of the left ventricle and the oxygen consumption of the myocardium are dependent on the size of the heart, i.e., heart weight. A larger mammal has a larger vasculature to perfuse. The larger amount of energy is derived from increased coronary blood flow, i.e., an increased myocardial oxygen supply. This results in a constant ratio of cardiac efficiency.

It has long been established in cardiac mechanics that it is more costly to perform pressure work than to perform volume work (Evans and Matsuoka, 1915). This stems from the fact that myocardial oxygen consumption is correlated to the tension-time index (Sarnoff et al., 1958) or the rate-pressure product, i.e., heart rate times systolic pressure. Since blood pressure remains relatively constant in mammals, the variations in oxygen consumption are therefore related to heart rate differences. In fact, the oxygen consumption per heart weight is proportional to heart rate, i.e., in a constant ratio to heart rate. Many attempts, other than Equation (9.2), have been made to correlate hemodynamic parameters with myocardial oxygen consumption. More recently, the pressure-volume area (PVA) has been shown to correlate with MVO_2, and this correlation is independent of heart rate. The pressure-volume loop is also termed the work loop, since the area under the loop is the EW as we have shown earlier. EW represents the external mechanical work performed by the left ventricle. PVA can be used in the computation of cardiac efficiency, i.e.,

$$e = EW/PVA \qquad (9.5)$$

Both PVA and EW have the units of energy. It should be cautioned, however, that PVA is only a quantitative correlate and not a direct measure of myocardial oxygen consumption.

The pressure and flow waveforms in the aortas of mammals are similar in shape. Although in the current discussion only a steady energy flow component was used in the computation, it should be clear that the pulsatile energy components are proportional to flow, as aortic pressure remains constant in mammals. Since aortic pressure also serves as the perfusion pressure to the coronary arteries, this constancy in pressure across mammalian species indicates that coronary flow is determined by its vascular resistances (R_c), which increase with decreasing body weight, that is, it is larger in smaller mammals, viz:

$$R_c = p/CBF \qquad (9.6)$$

Increases in heart rate with decreasing body weight appears to decrease the mechanical efficiency of the heart. But this is accompanied by a decrease in stroke volume and a decrease in oxygen consumption. Consequently, cardiac efficiency is again preserved. In the very small mammals, it has been shown that oxygen consumption is high, because of a higher heart weight to body weight ratio. This is in the case of the shrew, the smallest mammal, having a body weight of 3 to 5 g. It is not clear what physiological parameters actually set the constraints to cause deviations from the dimensional analysis and allometry. Some proposed that thermoneutrality demands that small mammals should have a higher metabolic rate. In Chapter 7, we have shown that metabolic rate is related to heart rate. This means a higher heart rate would be expected. This cannot be achieved, as the cardiac excitation conduction mechanism requires a finite amount of time. Consequently, the disproportional increase in oxygen demand and consumption in the very small mammals must be attributed to the large heart size, since a large increase in heart rate is not found.

During altered physiological states, cardiac efficiency is likely to change. In exercise, for instance, heart rate normally increases so as to increase cardiac output and thence, places greater oxygen demand on the heart. This tends to reduce the cardiac efficiency. This efficiency is also decreased in the diseased heart. The efficiency of a failing heart subjected to stress, such as mild to severe exercise, will decrease further. This is one of the underlying principles of the modern stress test for diagnosing the cardiac state in patients.

Allometry and dimensional analysis are not without their shortcomings, as discussed previously. Their careful use in combination with physical laws or hemodynamic principles can explain certain structural and functional correlates that arise from the natural design features. Questions that remain unanswered include why the mammalian heart works at a mechanical efficiency of only about 20% and whether this is related to functional optimality. If so, what factors govern such an optimal design feature?

9.2 GEOMETRY AND ELASTICITY: LOW LOSS AORTA AND BRANCHING CHARACTERISTICS

In studying optimality features of the vascular system, arterial branching has interested many investigators for some decades. Thompson (1917) touched on the subject in an interesting manner and cited Roux and Hess, who had proposed branching rules and derived a formula for the minimum angle of branching. Cohn (1954, 1955), in studying the optimum radius and length of daughter vessels in a bifurcation, derived a formula for the optimum radius of the branches. It is of interest to mention here that Murray (1926) had earlier found exactly the same solution and that the area ratio (AR) corresponds indeed to minimum reflection at a symmetrical bifurcation, as calculated by Karreman (1952), Womersley (1958), Zamir (1977), and Uylings (1977). Rosen (1967) gave an introduction on the subject of branching toward the view of optimality.

In general, the optimal angle of a bifurcation should be about 75°, as shown by Murray (1926) in some morphometrical data he obtained and theoretically calculated by Zamir (1976, 1977). However, additional data are necessary to confirm this. We have measured the branching angles at the aorto-iliac junction in man, dog, cat, and rabbit. The results of (72 ± 5) degrees come very close to Murray's finding.

It appears that the arterial system is so branched and tapered as to provide low impedance to flow to maintain the arterial system in a form sufficient to meet metabolic needs and to provide an optimal structure for pulse wave transmission.

Goetz et al. (1955, 1957, 1960) examined the circulation of the giraffe. They found the aortic pressure waveform to be similar to that of other mammals, but with a much higher mean blood pressure. The unusually long neck and carotid arteries make this mammalian species an exception in this regard.

We have shown that the arterial system exhibits branching characteristics. It has been speculated for some time that branching morphology may modify pulse-transmission characteristics. Experimental studies in mammals, however, have shown that impedances at proximal mother vessel and distal daughter branches at arterial branching junctions are essentially matched. The significance of this is that blood pressure pulse can propagate with minimum loss through these junctions (Li et al., 1984, 1985). Naturally, in diseased conditions, these junctional impedances can no longer be matched. This will result in an increased wave reflection of the forward propagating pulse and, consequently, increased pressure pulse amplitude. As a result, a higher shear stress is exerted on the lumen wall. Since elasticity increases with blood pressure, this will imply that the increased wall elasticity is a dominant factor in altering pulse transmission at the junction.

Geometric factors can be of considerable importance also. Narrowing of blood vessels can be more detrimental to pulse transmission than an increase in elastic properties (Li, 1985). We shall here demonstrate the importance of structure on function, namely, the relative importance of geometric and elastic properties in governing pulsatile transmission through vascular branching junctions. Quantitative relations involving elastic and geometric variables can be formulated to give an account of their relative contributions to pulse propagation and reflections at an arterial branching junction. Further illustration of the theoretical concept can be obtained from experimental data.

It is clear that the propagating pulse waveform is dependent on the elastic as well as the geometric properties of the artery. To analyze pulse transmission at a branching junction, characteristic impedances that incorporate both elastic and geometric properties can be formulated from linear transmission line theory. According to the water-hammer formula (Li et al., 1984), they are

$$Z_o = \rho c / \pi r^2 = \rho c / A \qquad (9.7)$$

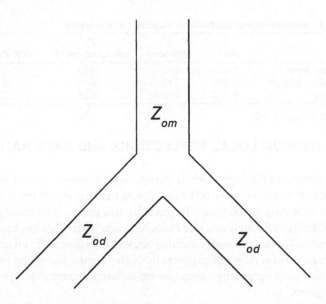

Figure 9.1 Branching vascular junction showing a mother vessel branched into two equal branches of daughter vessels (bifurcation).

where A is the cross-sectional area, and by substitution

$$Z_o = (Eh\rho/2r)^{1/2}/A \tag{9.8}$$

As an example, for a bifurcation with two daughter vessels of characteristic impedances Z_1 and Z_2,

$$1/Z_d = 1/Z_1 + 1/Z_2 \tag{9.9}$$

In the case of an equal bifurcation, the branching daughter vessels have identical properties (Figure 9.1),

$$Z_1 = Z_2$$

$$1/Z_d = 2/Z_1 \tag{9.10}$$

or for n equal daughter branches,

$$1/Z_d = n/Z_1 \tag{9.11}$$

As an example, Table 9.1 summarizes experimental data obtained in dog aorto-iliac trifurcation. The abdominal aorta branches into its continuation with characteristic impedance Z_2 and right and left iliac arteries each with characteristic impedance Z_1. Hence,

$$Z_m/Z_d = (2Z_m/Z_1) + Z_m/Z_2 \tag{9.12}$$

Table 9.1   Measured and Calculated Vascular Parameters

	A (cm^2)	c (cm/sec)	Z_o (dyn.sec.cm^{-5})	E (10^6 dyn/cm^2)
Abdominal aorta	0.415	660	1686	5.54
Continuation branch	0.205	710	3671	6.41
Iliac artery	0.115	765	7051	7.44

Note: Assuming r/h = 6.

9.3 MINIMUM LOCAL REFLECTIONS AND AREA RATIOS

By examination of the characteristic impedances of mother and daughter vessels, one can conclude that pulse wave reflection due to vascular branching is minimal (Li, 1984). Since reflection is energetically wasteful, this means little energy is lost due to pulse transmission through vascular branching junctions. This has been attributed to close to optimal area ratios and branching angles. Also geometric effect rather than elastic effect dominates pulse propagation through vascular branching junctions (Li, 1985). Whether such optimality exists among different mammalian species has not been investigated.

9.3.1 THE AREA RATIO CONCEPT

The concept of area ratio has been utilized in studying wave reflections at the arterial junctions. An area ratio (the ratio of the sum of the areas of daughter vessels to that of the mother vessel) for a reflectionless bifurcation of about 1.15. For optimum energy transfer, the area ratio needs to be close to one.

The area ratio concept has recently received renewed attention. Its practical applications have been demonstrated by, for instance, Gosling et al. (1971). Area ratio is normally measured by injecting radio-opaque dye to obtain radiographs and by assuming a circular cross section for the junction vessels.

In small, muscular vessels, viscous damping is appreciably more important than in large vessels. In these vessels, the point of minimum reflection (in the reflection vs. area ratio plot) is shifted to a larger area ratio, accompanied by a larger phase change.

The fraction of the pressure pulse that is reflected at the junction due to un-matched branching vessel characteristic impedances is

$$\Gamma_v = \left(Z_d - Z_m\right)/\left(Z_d + Z_m\right) = \left(1 - Z_m/Z_d\right)/\left(1 + Z_m/Z_d\right) \qquad (9.13)$$

where Z_m and Z_d are the characteristic impedances of the mother and daughter vessels, respectively.

This latter leads to a junctional reflection of

$$\Gamma_v = \left(1 - n Z_m/Z_1\right)/\left(1 + n Z_m/Z_1\right) \qquad (9.14)$$

Let

$$Z_m = \rho c_m/A_m \quad \text{and} \quad Z_1 = \rho c_1/A_1 \qquad (9.15)$$

then, for n equal branches, we obtain

$$Z_m/Z_d = \left(n A_1/A_m\right)\left(c_m/c_1\right) \tag{9.16}$$

where

$$AR = n A_1/A_m \tag{9.17}$$

is the area ratio (AR) and c_m/c_1 is the velocity ratio,

$$c_m/c_1 = \left(E_m/E_1\right)^{1/2} \tag{9.18}$$

This assumes that the h/r ratio is relatively constant for branching vessels (Li et al., 1981).

Since

$$AR = n\left(r_1/r_m\right)^2 \tag{9.19}$$

it is clear that alterations in branching vessel lumen radii could exert a more significant effect on Γ_v than changes in elastic moduli.

For constant elasticity $c_m/c_1 = 1$ and Γ_v becomes dependent only on area ratio, i.e.,

$$\Gamma_v = (1 - AR)/(1 + AR) \tag{9.20}$$

We shall illustrate below that branching geometry has a more dominant effect than vessel elasticity on pulse reflection at a vascular junction.

The changes in pulse reflection as a function of percentage decrease in radius of either one or both iliac arteries are plotted in Figure 9.2. A 100% reduction in radius represents total vessel occlusion. For instance, a 50% lumen narrowing of the left iliac artery will increase pulse reflection from the normal control value of 3 to 14%; the value would increase to 27% if both iliac arteries were partially occluded to the same extent. A 50% increase in stiffness (E) alone in the left iliac artery increases pulse reflection only from 3 to 8%. Hence, geometric change produces a more significant effect than elastic effect on pulse propagation at the junction. This difference is more pronounced when a large vessel is involved. Figure 9.3 shows that a 50% reduction in radius of the abdominal aorta could result in a pulse reflection of 58%. While an increase in stiffness by 50% produces a corresponding junctional reflection of only 7%.

9.4 THE NATURAL DESIGN CHARACTERISTICS

Mammals, differing in appearance and body size, have cardiovascular systems that exhibit similar functions. The design characteristics are particularly and amazingly similar. The basic units or the fundamental building blocks are of similar constituents, size, and operating function. A larger mammal simply has a larger heart,

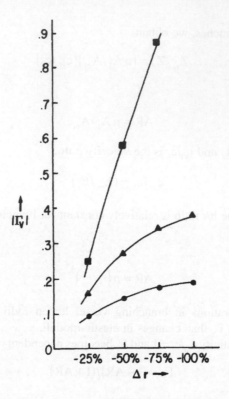

Figure 9.2   Pulse wave reflections at vascular junction when geometry is altered. (From Li, J. K-J., *Bull. Math. Biol.*, 48:97–103, 1986. With permission.)

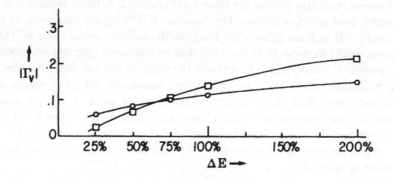

Figure 9.3   Pulse wave reflections at vascular junction when elastic properties are altered. (From Li, J. K-J., *Bull. Math. Biol.*, 48:97–103, 1986. With permission.)

with a larger number of constituent muscle fibers or sarcomeres. These sarcomeres, constituting the myocardium, exercise their changes in lengths to provide tension that is necessary for pressure development in order to propel blood to the vascular system. As a muscular pump, it needs a supply of energy. At about 20%, the efficiency of the mammalian cardiac pump does not differ. Although the mechanical efficiency is considerably higher, the ability of the heart to generate work is similar among mammals. The amount of work that the heart performs is sufficient to meet the vascular demands. Thus, the work is performed to overcome the load in order to perfuse the vascular beds. The mechanical design includes features that manifest in the compliance of the aorta, the characteristic impedance of the aorta, and the peripheral resistance.

The aorta's compliance increases in a larger mammal to accommodate the larger stroke volume ejected by a larger heart. The peripheral resistance, closer to the heart as in smaller mammals, increases. This adjustment maintains an inverse relationship to cardiac output. The pulse waveforms at corresponding arterial sites are similar. The increased elastic stiffness, and hence, pulse wave velocity toward the periphery, is increased to facilitate pulse transmission. But the system's natural design is more marvelous at vascular branching junctions. The wave reflections are minimal here, since branching vessel impedances are practically matched. Thus, pulse wave can be transmitted at utmost efficiency through these junctions with minimal losses. To prevent greater reflected waves from reaching the heart, hence reducing the ventricular ejected stroke volume, these vascular branching junctions present an obstacle - greater unmatching for the returning waves.

Perhaps, the most important functional design feature is the heart rate. Is the number of heart beats fixed in the life span of a mammal? The answer, seemingly, is positive. A decreasing heart size, hence, body size, associates with an increasing heart rate and the metabolic turnover rate. But the metabolic turnover rate per unit heart rate remains the same. Heart rate is adjusted to meet functional demands. Dynamic features provide rapid changes, while structural design features alter more slowly to adapt to functional demands. Such integral structural-functional interplay optimizes the performance of the mammalian cardiovascular system.

References

Adolph, E.F., Quantitative relations in the physiological constitutions of mammals, *Science*, 109:579, 1949.

Adolph, E.F., Physiological integrations in action, *Physiologist*, 25(2), Suppl. 1982.

Aperia, A., Hemodynamic studies, *Skand. Arch. Physiol.*, 83, Suppl. 16:1-230, 1940.

Attinger, E.O., A. Anne, and D.A. McDonald, Use of Fourier series for the analysis of biological systems, *Biophys. J.*, 6:291-304, 1966.

Attinger, E.O., H. Sugawara, A, Navarro, A. Riccceto, and R. Martin, Pressure-flow relations in dog arteries, *Circ. Res.*, 19:230-246, 1966.

Avolio, A.P., M.F. O'Rourke, K. Mang, and P.T. Bason, A comparative study of pulsatile arterial hemodynamics in rabbits and guinea pigs, *Am. J. Physiol.*, 230:868-875, 1976.

Berger, D.S. and J.K.-J. Li, Concurrent compliance reduction and increased peripheral resistance in the manifestation of isolated systolic hypertension, *Am. J. Cardiol.*, 65:67-71, 1990.

Braunwald, E., J. Ross, Jr., and E.H. Sonnenblick, *Mechanisms of Contraction of the Normal and Failing Heart*, Little Brown, Boston, 1976.

Brody, S., *Bioenergetics and Growth*, Reinhold, New York, 1945.

Buckingham, E., On physically similar systems: illustrations of the use of dimensional equations, *Phys. Rev.*, 4:345, 1915.

Burton, A.C., Relation of structure to function of the tissues of walls of blood vessels, *Physiol. Rev.*, 34:619-642, 1954.

Calder, W.A., III, Scaling of physiological processes in homeothermic animals, *Annu. Rev. Physiol.*, 43:301, 1981.

Caro, C.G., J.G. Pedley, R.C. Schroter and W.A. Seed, *The Mechanics of the Circulation*, Oxford University Press, New York, 1978.

Clark, A.J., *Comparative Physiology of the Heart*, Macmillan, New York, 1927.

Cohn, D., Optimal systems. I. The vascular system, *Bull. Math. Biophys.*, 16:59, 1954.

Cohn, D., Optimal systems. II, *Bull. Math. Biophys.*, 17:219, 1955.

Cox, R.H., Pressure dependence of the mechanical properties of arteries *in vivo*, *Am. J. Physiol.*, 229:371, 1975.

DuBois D. and E.F. DuBois, A formula to estimate the approximate surface area if height and weight be known, *Arch. Intern. Med.*, 17:863-871, 1916.

Evans, C.L. and Y. Matsuoka, The effect of various mechanical conditions on the gaseous metabolism and efficiency of the mammalian heart, *J. Physiol.*, 49:378-405, 1915.

Fahraeus, R. and T. Lindquist, The viscosity of blood in narrow capillary tubes, *Am. J. Physiol.*, 96:562, 1931.

Fich, S. and J.K.-J. Li, Aorto-ventricular dynamics: theories, experiments and instrumentation, *CRC Crit. Rev. Biomed. Eng.*, 9:245, 1983.

Foung, P.L.-G. and J.K.-J. Li, Spectral analysis of heart rate in beta-blockade patients, Proc. 38th Annu. Conf. Eng. Med. Biol., 27:127, 1985.

Frank, O., Die Grundform des arteriellen Puls, Z. Biol., 37:483, 1899.

Galilei, G., *Dialogues Concerning Two New Sciences*, 1638, Dover, ed., New York, 1914.

Geipel, P.S. and J. K-J. Li, Nitroprusside abolishes increased arterial wave reflections from methoxamine-induced hypertension, Proc. 15th NE Bioeng. Conf., pp. 131-132, 1989.

Goetz, R.H. and O. Budtz-Olsen, Scientific safari: circulation of the giraffe, *S. African Med. J.*, 29:773, 1955.

Goetz, R.H. and E.N. Keen, Some aspects of the cardiovascular system of the giraffe, *Angiology*, 8:542, 1957.

Goetz, R.H., J.V. Warren, O.H. Gauer, J.L. Patterson, Jr., J.T. Doyle, E.N. Keen, and M. McGregor, Circulation of the giraffe, *Circ. Res.*, 8:1049-1058, 1960.

Gordon, A.M., A.F. Huxley and F.J. Julian, The variation in isometric tension with sacromere length in vertebrate muscle fibers, *J. Physiol. (Lond.)*, 184:170-192, 1966.

Gosling, R.G., D.L. Newman, N.L.R. Bowden, K.W. Twinn, Aortic configuration and pulse wave reflection, *Br. J. Radiol.*, 44:850, 1971.

Gow, B.S. and M.F. O'Rourke, Comparison of pressure and flow in the ascending aorta of different mammals, *Proc. Austral. Physiol. Pharm. Soc.*, 1:68, 1970.

Green, H.D., Circulatory system: physical principles, in *Medical Physics 2*, Glasser, O., Ed., Year Book Publishers, New York, 1950.

Gunther, B., Allometric ratios, invariant numbers and the theory of biological similarity, *Physiol. Rev.*, 55:659, 1975.

Gunther, B. and E. Guerra, Biological similarities, *Acta Physiol. Lat. Am.*, 5:169, 1955.

Gunther, B. and L. De La Barra, Physiometry of the mammalian circulatory system, *Acta Physiol. Lat. Am.*, 16:32, 1966.

Gunther, B. and L. De La Barra, Theories of biological similarities, non-dimensional parameters and invariant numbers, *Bull. Math. Biophys.*, 28:9-102, 1966.

Guyton, A.C. and T.G. Coleman, Long term regulation of the circulation, in *Physical Bases of Circulatory Transport*, Reeve, E.B. and A.C. Guyton, Eds., Saunders, Philadelphia, 1967.

Hales, S., *Statical Essays Containing Haemostaticks*, Innys and Manby, London, 1733.

Hamilton, W.F. and P. Dow, An experimental study of the standing waves in the pulse propagation through the aorta, *Am. J. Physiol.*, 125:48, 1939.

Harvey, P.H., On rethinking allometry, *J. Theor. Biol.*, 95:37-41, 1982.

Heusner, A.A., Energy metabolism and body size, *Respir. Physiol.*, 48:1, 1982.

Heusner, A.A., Body size, energy metabolism, and the lungs, *J. Appl. Physiol.*, 54:867-873, 1983.

Heusner, A.A., Biological similitude: statistical and functional relationships in comparative physiology, *Am. J. Physiol.*, 246:R839-R845, 1984.

Hill, A.V., The series elastic component of muscle, *Proc. Roy. Soc. London Ser. B*, 137:273-280, 1950.

Holt, J.P., E.A. Rhode, W.W. Holt, and H. Kines, Geometric similarity of aorta, venae cavae, and certain of their branches in mammals, *Am. J. Physiol.*, 241:R100, 1981.

Holt, J.P., E.A. Rhode, and H. Kines, Ventricular volumes and body weights in mammals, *Am. J. Physiol.*, 215:704, 1968.

Hossdorf, H., *Model Analysis of Structures*, Van Nostrand Reinhold, New York, 1974.

Huxley, H.E., The contraction of muscle, *Sci. Am.,* 199:66-82, 1958.

Huxley, J.S., *Problems of Relative Growth*, Methuen, London, 1932.

Iantorno, S. and J.K.-J. Li, Compliance indices in the assessment of cardiac diseases, Proc. Int. Conf. Eng. Med. Biol., 10:247-248, 1988.

Iberall, A.S., Growth, form and function in mammals, *Annu. N.Y. Acad. Sci.*, 231:77, 1974.

Iberall, A.S., Some comparative scale factors for mammals, *Am. J. Physiol.*, 237:R7, 1979.

Intaglietta, M., R.F. Parvula, and W.R. Thompkins, Pressure measurement in the mammalian microvasculature, *Microvasc. Res.*, 2:212-220, 1970.

Juznic, G. and H. Klensch, Vergleichende physiologische untersuchunger uber das verhalten der indices fur energieaufwand und leistung des herzens, *Arch. ges Physiol.*, 280:3845, 1964.

Kamiya, A., T. Togawa and A. Yamamoto, Theoretical relationship between the optimal models of the vascular tree, *Bull. Math. Biol.*, 36:311, 1974.

Karreman, G., Some contributions to the mathematical biology of blood circulation, *Bull. Math. Biophys.*, 14:327, 1952.

Kenner, T., Flow and pressure in arteries, in *Biomechanics*, Fung, Y.C., N. Perroue, and M. Anliker, Eds., Prentice-Hall, Englewood Cliffs, NJ, 1972.

Kerkhof, P.L.M., End-Systolic Volume and the Evaluation of Cardiac Pump Function, Ph.D. dissertation, University of Utrecht, 1981.

Kleiber, M., Body size and metabolic rate, *Physiol. Rev.,* 27: 511-541, 1947.

Kleiber, M., *The Fire of Life*, John Wiley & Sons, New York, 1961.

Kleiber, M., Metabolic turn-over rate: a physiological meaning of the metabolic rate per unit body weight, *J. Theor. Biol.*, 53:199, 1975.

Krovetz, L.J., The physiological significance of body surface area, *J. Pediatr.*, 67:841-862, 1965.

Krovetz, L.J., The effect of vessel branching on hemodynamic stability, *Phys. Med. Biol.*, 10:417, 1965.

Lambert, R. and G. Teissier, Theorie de la similitude biologique, *Annu. Physiol. Physiocochem. Biol.*, 3:212, 1927.

Lanoce, V.M., Similarity Analysis of the Mammalian Cardiovascular System, M.S. thesis, Rutgers University, New Brunswick, NJ, 1985.

Li, J.K.-J. Mammalian Hemodynamics: Wave Transmission Characteristics and Similarity Analysis, Ph.D. dissertation, University of Pennsylvania, Philadelphia. 1978. University Microfilms, Ann Arbor, MI.

Li, J.K.-J., Similarity criteria and modelling in mammalian hemodynamics, Proc. 144th Am. Assoc. Adv. Sci. Natl. Meet., 144:119, 1978.

Li, J.K.-J., Cardiovascular diagnostic parameters derived from pressure and flow pulses, in *Frontiers of Engineering in Health Care*, Vol. 4, 186-189, 1982.

Li, J.K.-J., A new similarity principle for cardiac energetics, *Bull. Math. Biol.*, 45:1005-1011, 1983.

Li, J.K.-J., Hemodynamic significance of metabolic turnover rate, *J. Theor. Biol.*, 103:333-338, 1983.

Li, J.K.-J., Dimensional analysis: application to mammalian cardiac energetics, Proc. 149th Am. Assoc. Adv. Sci. Natl. Meet., 149, 1983.

Li, J.K.-J., Similarity analysis of mammalian cardiac energetics, *Physiologist*, 26:A25, 1983.

Li, J.K.-J., Comparative hemodynamics: Laplace's law applied to the heart, Proc. 150th Am. Assoc. Adv. Sci. Natl. Meet., 1984.

Li, J.K.-J., Pulse wave reflections at the aortic-iliac junction, *Angiology*, 36:516-521, 1985.

Li, J.K.-J., Dominance of geometric over elastic factors in pulse transmission through arterial branching, *Bull. Math. Biol.*, 48:97-103, 1986.

Li, J.K.-J., Comparative cardiac mechanics: Laplace's law, *J. Theor. Biol.*, 118:339-343, 1986.

Li, J.K.-J., Time domain resolution of forward and reflected waves in the aorta, *IEEE Trans. Biomed. Eng.*, BME-33:783-785, 1986.

Li, J.K.-J., Regional left ventricular mechanics during myocardial ischemia, in *Simulation and Modeling of the Cardiac System*, Sideman, S., Ed., Martinus Nijhoff, The Hague, The Netherlands, 1987, pp. 451-462.

Li, J.K.-J., *Arterial System Dynamics*, New York University Press, New York, 1987.

Li, J.K.-J., Laminar and turbulent flow in the mammalian aorta: reynolds number, *J. Theor. Biol.*, 135:409-414, 1988.

Li, J.K.-J., Increased arterial pulse wave reflections and pulsatile energy loss in acute hypertension, *Angiol. J. Vasc. Dis.*, 40:730-735, 1989.

Li, J.K.-J., Dynamics of left ventricle-arterial system interaction, in, *Imaging, Measurement and Analysis of the Heart*, S. Sideman, S. and R. Beyar, Eds., Hemisphere Publishing, New York, 1990.

Li, J.K-J. Feedback effects in heart-arterial system interaction, interactive phenomena in the cardiac system, *Adv. Exp. Med. Biol.*, 346:325-333, 1993.

Li, J.K.-J., K. B. Campbell and A. Noordergraaf, Similarity principle of arterial dynamics in mammals, Proc. 30th Annu. Conf. Eng. Med. Biol., 19:329, 1977.

Li, J.K.-J., K.B. Campbell and A. Noordergraaf, Design criteria for mammalian arterial trees, *Fed. Proc.*, 37:218, 1978.

Li, J.K.-J., T. Cui, and G. Drzewiecki, A nonlinear model of the arterial system incorporating a pressure-dependent compliance, *IEEE Trans. Biomed. Eng.*, BME-37:673-678, 1990.

Li, J.K.-J., P.S. Geipel, D.S. Berger, C.P. Falkenhagen, and G. Drzewiecki, Model-based parameters and indices for assessing the cardiovascular state, Proc. 11th Int. Conf. Eng. Med. Biol., 11:114-115, 1989.

Li, J.K.-J., V.M. Lanoce, G. M. Drzewiecki, P.S. Geipel, D.S. Berger, and C.P. Falkenhagen, Cardiovascular parameters normalization by Huxley's allometric equation, *Fed. Am. Soc. Exp. Biol. J.*, 1990.

Li, J.K.-J., J. Melbin and A. Noordergraaf, Pulse transmission to vascular beds, Proc. 33rd Annu. Conf. Eng. Med. Biol., 22:86, 1980.

Li, J.K.-J., J. Melbin and A. Noordergraaf, Functional design: heart rate, Proc. 35th Annu. Conf. Eng. Med. Biol., 24:35, 1982.

Li, J.K.-J., J. Melbin, and A. Noordergraaf, Directional disparity of pulse wave reflections in dog arteries, *Am. J. Physiol.*, 247, H95-H99, 1984.

Li, J.K.-J., J. Melbin, R.A. Riffle and A. Noordergraaf, Pulse wave propagation, *Circ. Res.*, 49:442-452, 1981.

Li, J.K-J. and A. Noordergraaf, Similar pressure pulse propagation and reflection characteristics in aortas of mammals, *Am. J. Physiol.*, 261:R519-521, 1991.

Li, J.K-J. and Y. Zhu, Arterial compliance and its pressure-dependence in hypertension and vasodilation, *Angiol. J. Vasc. Dis.*, 45:113-117, 1994.

Little, R.C., *Physiology of the Heart and Circulation*, Year Book Medical Publishers, Chicago, 1985.

Loiselle, D.S. and C.L. Gibbs, Species differences in cardiac energies, *Am. J. Physiol.*, 490-498, 1979.

Longmore, D., *The Heart*, McGraw-Hill, New York, 1971.

Martin, R.R.and H. Haines, Application of Laplace's law to mammalian hearts, *Physiology*, 34:959, 1970.

McDonald, D.A., *Blood Flow in Arteries*, 1st ed., Arnolds, London, 1960; 1974, 2nd ed.

McMahon, T., Size and shape in biology, *Science*, 179:1201-1204, 1973.

Murray, C.D., The physiological principle of minimum work applied to the angle of branching of arteries, *J. Gen. Physiol.*, 9:835, 1926.

Murray, C.D., The physiological principle of minimum work, *Proc. Natl. Acad. Sci.*, 12:207, 1926.

Newton, I., *Philosophiae Naturalis Principia Mathematica*, Cantabrigiae, 1735. Lib. sec. VII, Propositio 32:294.

Noordergraaf, A., *Circulatory System Dynamics*, Academic Press, 1978.

Noordergraaf, A., J.K.-J. Li, and K.B. Campbell, Mammalian hemodynamics: a new similarity principle, *J. Theor. Biol.*, 79:485, 1979.

Oka, S., *Cardiovascular Hemorheology*, Cambridge University Press, New York, 1981.

Pasquis, A., A. Lacaisse, and P. Dejeurs, Maximal oxygen uptake in four species of small mammals, *Respir. Physiol.*, 9:298-309, 1970.

Patterson, S.M., M.T. Ortiz, D. G. Daut, and J.K.-J. Li, Noninvasive diagnosis of left ventricular hypertrophy, Proc. 39th Annu. Conf. Eng. Med. Biol., 28:370, 1986.

Poiseuille, J.L.M., Ann. Sci. Nat. Ser.2, *Zoology*, 5:111, 1836.

Porje, I.G., Studies of the arterial pulse wave, particularly in the aorta, *Acta Physiol. Scand.*, 13, Suppl. 42, 1946.

Porje, I.G., The energy design of the human circulatory system, *Cardiology*, 51:293, 1967.

Puri, N.N., J.K.-J. Li, S. Fich, and W. Welkowitz, Control system for circulatory assist devices: determination of suitable control variables, *Trans. Am. Soc. Artif. Intern. Organs*, 28:127-132, 1982.

Reynolds, O., An experimental investigations of the circumstances which determine whether the motion of water shall be direct or sinuous and of the law of resistance in parallel channels, *Philos. Trans. Roy. Soc. London*, 174:935, 1883.

Richards, R.L., *Peripheral Arterial Diseases*, Livingstone, London, 1970.

Robard, S., F. Williams, and C. Williams, The spherical dynamics of the heart, *Am. Heart J.*, 57:348-360, 1959.

Robard, S., The burden of resistance vessels, *Circ. Res.*, 28:Suppl. 12, 1971.

Rosen, R., The vascular system, in *Optimality Principles in Biology*, Butterworths, London, 1967.

Rosen, R., Dynamical similarity and the theory of biological transformations, *Bull. Math. Biol.*, 49:549, 1978.

Rosen, R., Role of similarity principles in data extrapolation, *Am. J. Physiol.*, 244:R591, 1983.

Rushmer, R.F., *Structure and Function of the Cardiovascular System*, W.B. Saunders, Philadelphia, 1972.

Sagawa, K., The ventricular pressure-volume diagram revisited, *Circulation*, 433:677, 1978.

Salotto, A.G., L.F. Muscarella, J. Melbin, J.K.-J. Li, and A. Noordergraaf, Pressure pulse transmission into vascular beds, *Microvasc. Res.*, 32:152-163, 1986.

Sarnoff, J.S., and E. Berglund, Ventricular function: I. Starling's law of the heart studied by means of simultaneous right and left ventricular function curves in the dog, *Circulation*, 9:706-718, 1954.

Sarnoff, J.S., E. Brunwald, G.H. Welch, Jr., R.B. Case, W.N. Stainsby, and R. Macruz, Hemodynamic determinants of oxygen consumption of the heart with special reference to the tension-time index, *Am. J. Physiol.*, 192:148-156, 1958.

Savaqeau, M.A., Growth equations: a general equation and a survey of special cases, *Math. Biosci.*, 48:267-278, 1980.

Schmidt-Nielsen, K., Energy metabolism, body size, and problems of scaling, *Fed. Proc. Fed. Am Soc. Exp. Biol.*, 29:1524-1532, 1970.

Schmidt-Nielsen, K., *How Animals Work*, Cambridge University Press, 1972.

Schmidt-Nielsen, K., Scaling in biology: The consequences of body size, *J. Exp. Zool.*, 194:287, 1975.

Schmidt-Nielsen, K., *Scaling: Why is Animal Size so Important?*, Cambridge University Press, 1984.

Smith, R.J. Rethinking allometry, *J. Theor. Biol.*, 87:97, 1980.

Somlyo, A.P. and A.V. Somlyo, Vascular smooth muscle. I. Normal structure, pathology, biochemistry and biophysics, *Pharm. Rev.*, 20:197-272, 1968.

Sonnenblick, E.H., D. Spiro and H.M. Spotnitz, Ultrastructural basis of Starlins's law of the heart: the role of the sacromere in determining ventricular size and stroke volume, *Am. Heart J.*, 68:336-346, 1964.

Stahl, W.R., Similarity and dimensional biology, *Science*, 137:205, 1962.

Stahl, W.R. Similarity analysis of biological systems, *Persp. Biol. Med.*, 6:291, 1963.

Stahl, W.R. The analysis of biological similarity, *Adv. Biol. Med. Phys.*, 9:356, 1963.

Stahl, W.R., Organ weights in primates and other mammals, *Science*, 150:1039-1042, 1965.

Stahl, W.R., Scaling of respiratory variable in mammals, *J. Appl. Physiol.*, 22:453-460, 1967.

Starling, E.H. and M.B. Visscher, The regulation of the energy output of the heart, *J. Physiol.*, 62:243-261, 1926.

Starling, E.H., *Linacre Lecture on the Law of the Heart*, London, 1918.

Taylor, C.R., Structural and functional limits to oxidative metabolism: insights from scaling, *Annu. Rev. Physiol.*, 49:135-146, 1987.

Thompson, D.W., *On Growth and Form*, Cambridge University Press, 1917.

Uylings, H.B.M., Optimization of diameters and bifurcation angles in lung and vascular tree structures, *Bull. Math. Biol.*, 39:509, 1977.

White, L., H. Haines and T. Adams, Cardiac output related to body weights in small mammals, *Comp. Biochem. Physiol.*, 27:559-565, 1968.

Womersley, J.R., Oscillatory flow in arteries: the reflection of the pulse wave at junctions and rigid inserts in the arterial system, *Phys. Med. Biol.*, 2:213, 1958.

Woods, R.H., A few applications of a physical theorem to membranes in the human body in a state of tension, *J. Anat. Physiol.*, 26:362-370, 1892.

Zhu, Y., Hemodynamic Basis and Nonlinear Model Analysis of Hypertension and Aging, M.S. thesis, Rutgers University, Piscataway, NJ, 1993.

Yates, F.E., Comparative physiology: compared to what?, *Am. J. Physiol.*, 237: R1, 1979.

Yates, F.E., Comparative physiology of energy production: homeotherms and poikliotherms, *Am. J. Physiol.*, 241:R1, 1981.

Young, D.F. and N.R. Cholvin, Application of the concept of similitude to pulsatile blood flow in arteries, *ASME Biomed. Fluid Mech. Symp.*, 1966.

Zamir, M., Optimality principles in arterial branching, *J. Theor. Biol.*, 62:227, 1976.

Zamir, M., Shear forces and blood vessel radii in the cardiovascular system, *J. Gen. Physiol.*, 69:449, 1977.

Zweifach, B.W., Quantitative studies of microcirculatory structure and function, *Circ. Res.*, 34:858-866, 1974.

Index

A

Actin, 59
Adventitia, 18
Afterload, 71, 98, See also Input impedance
Allometric equations, 24-27, 41
  blood flow fluid dynamics, 101-105
  cardiac parameters, 33, 114-115
  heart rate, 114-115
  mathematics, 41-45
    linear equation solutions, 48
    matrices, 45-48
  metabolic rate, 120
Allometry, See also Body weight; Simarity
     analysis; specific parameters
  cardiovascular control, 126-127
  circulatory system, 32-35
  creating new simarity criteria, 56
  defined, 25
  hemodynamics, 35-39
  mammalian heart function, 67-69
  simple, deviations from, 30-31
Aorta, 15, See also Arterial system
  compliance, 128-130
  diameter/length ratio, 27
Area ratio, 16-18, 140, 142-143
Arterial blood pressure, See Blood pressure
Arterial system
  anatomical and structural organization, 15-18
  blood flow properties, 80-82
  branching, See Branching
  compliance, 105, 107, 128-130
  elastance, 129-132
  elasticity, 75-78, 132
  fluid dynamics, see Hemodynamics
  input impedance, 85-90, 98
  viscoelastic properties, 78-80
  wall components, 18, 75
  waveforms, See Pressure and flow waveforms
  windkessel model, 89
Arterioles, 21, 82, 118-120, 135-136

Arteriovenous caplaries, 21
Atria, 11-13

B

Basal metabolism, 27
Bernoulli principle, 82
Beta adrenergic blockers, 133, 136
Biological control systems, 125-127, See also
     Cardiovascular control
Biscuspid valves, 19
Blood, 4-7
  as non-Newtonian fluid, 120
  flow, See Hemodynamics
Blood pressure, 110
  allometric equation, 115
  dimensional matrix, 54
  hypertension, 134-136
  measurement, 53
  hypertension, 134-136
  regulation, 125
Blood pressure and flow waveforms, See
     Pressure and flow waveforms
Body size
  heart rate and, 1, 114
  organ size and, 31-32
  red cell size and, 22
Body surface area, 31, 115
Body weight
  allometric formula, 25
  arterial pulse wave velocity and, 92
  blood volume and, 6
  external cardiac work and, 116-117
  heart rate and, 29, 139
  heart weight and, 24, 36
  metabolic rate and, 117, 120
  velocity fluctuation ratio and, 104
Branching, 109, 139-142
  area ratio and, 142-143
  local reflection coefficient, 94
  wave reflections, 140, 142-143, 145

153